Fine WoodWorking®
on More Proven Shop Tips

Fine WoodWorking

on More Proven Shop Tips

Selections from
Methods of Work,
edited and drawn by
Jim Richey

The Taunton Press

Cover art by Jim Richey

First printing: February 1990
Second printing: October 1990
Third printing: January 1993
International Standard Book Number: 0-942391-43-8
Library of Congress Catalog Card Number: 89-40573
Printed in the United States of America

A FINE WOODWORKING Book

FINE WOODWORKING is a trademark of The Taunton Press, Inc.,
registered in the U.S. Patent and Trademark Office.

The Taunton Press, Inc.
63 South Main Street
Box 5506
Newtown, Connecticut 06470-5506

Introduction

In the month-to-month business of pulling together material for a magazine column, there's not much room for reflection. Like Charlie Chaplin's frantic cake decorator in the silent movie, we have barely enough time to add a flourish to each wonderful little idea before it falls off the assembly line into the printing press. That's why it is such a pleasure to step away for a while to gather together the last five years' worth of tips and tricks, and to see what shape the pieces take when they're put together.

This book is the second compilation of woodworking tips taken from the Methods of Work column in *Fine Woodworking* magazine. The first book, *Fine Woodworking on Proven Shop Tips,* covered material published through November of 1984, and this book picks up from there. Many of the tips in this second book focus on three of the shop's mainstays: the tablesaw, lathe and router. Of course, all the other tools are well represented, but there's something about a tablesaw that is more inspirational than, say, a radial-arm saw. The simple explanation might be that more woodworkers have these tools in their shops. My own theory is that the comparative simplicity and directness of these tools somehow amplifies the creative process. Whatever the reason, if you have a tablesaw, a lathe and a router in your shop, here you will find dozens of unexpected and new ways to use them.

This second book also contains an abundance of information on shop aids and construction tips, including shop vacuum systems. Some of the ideas in these pages are truly clever, like John Loughrey's shop-organizing cleat system and Jeffrey Gyving's pneumatic edging clamps.

This work is very much a collaboration. Jim Cummins, Jim Boesel, Roland Wolf and the rest of the staff at *Fine Woodworking* deserve special thanks for their talented help. Most of all, however, I would like to thank the 300 woodworkers who are the real authors of this book. It was their creativity and generosity in hatching these wonderful ideas that was the hard part. Making a book from them was easy.

Jim Richey, editor, Methods of Work

Contents

Benches and Vises

Chapter 1

Two sawhorses

Here's a sawhorse with no metal parts to mar your work. These horses stack neatly, and they can also be knocked apart easily for storage or transport. The plywood legs are 8-in. wide at the top, though 6 in. will work if you want to cut down on weight. The only caution is that the sliding joints must be cut tight enough so that they must be driven with a hammer, or the horse will wobble. This construction is much stronger than it looks at first glance—I've put one sawhorse through the teen-age student torture-test in my shop class, where it survives unbroken.
—*Mark Blieske, Winnipeg, Man.*

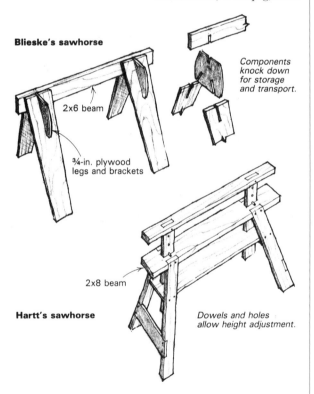

Blieske's sawhorse

2x6 beam

Components knock down for storage and transport.

¾-in. plywood legs and brackets

2x8 beam

Hartt's sawhorse

Dowels and holes allow height adjustment.

The adjustable-horse design shown above was originally published as a project in a 1958 Deltagram. I have put together several sets of these horses and find them to be valuable additions to any shop. With the extension down or removed, it is a sturdy sawhorse. With the extensions raised, two horses and a sheet of plywood make a handy layout table. The extensions can also be used to support long stock or plywood sheets when ripping on the table saw. —*Grover Hartt Jr., Dallas, Tex.*

Folding saw rack

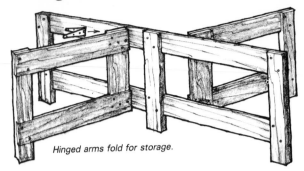

Hinged arms fold for storage.

This multipurpose folding rack takes the place of several sawhorses, yet when stored it occupies less space than one. Unfolded, it can support a 4x8 sheet of plywood for ripping or crosscutting. It's also handy for cutting 2x4s to length and other framing work. With a piece of plywood on top, it becomes a handy work platform. —*Phil Mackie, Rhinelander, Wis.*

Stacking sawhorses

When my husband, John, and I built our sawhorses, we took a lot of Sam Allen's good advice on the subject *(FWW #24)*. But we found that by modifying his basic design slightly the horses were much easier to deal with around our shop, a place that always seems a bit too crowded. To allow the horses to be stacked for storage, we notched and beveled the plates of ½-in. Baltic birch plywood that reinforce the legs. Only a touch of clearance is required for a comfortable and stable fit. Our horses stack six or eight high with no wobble.
—*Carolyn Grew-Sheridan, San Francisco, Calif.*

Bar-clamp gluing rack

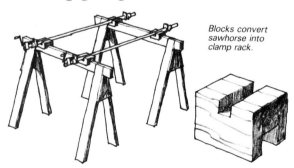

Blocks convert sawhorse into clamp rack.

I've noticed that many professional shops have clamp racks that are dedicated to edge gluing. These racks hold the bar clamps in position, allow the woodworker to inspect the underside of the boards for fit and free up the workbench for other operations.

Since dedicated racks may not be practical for a small shop, here's a simple clamp-holder block that converts a sawhorse to a clamp rack. The fixture is simply a short block of wood with a channel ripped in the bottom to ride the horse, and a slot crosscut in the top to fit the bar clamp. If the top of your horse is a 2x4, you may wish to construct the block by gluing up a sandwich, as shown in the sketch. Make as many pairs of blocks as you'll need. The blocks stabilize the bar clamps and provide full flexibility of clamp spacing.

—*George A. Burman, Fort Bragg, Calif.*

Trailer-ball power arm

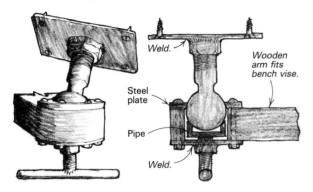

Weld.

Steel plate

Pipe

Weld.

Wooden arm fits bench vise.

This "poor man's power arm" is invaluable for carvers and sculptors because it lets you swivel and lock a workpiece at any convenient angle. The heart of the fixture is an old trailer ball. Although it certainly isn't necessary, I cut away the shoulder and narrowed the neck of the ball to allow a little more articulation of the joint. The ball rotates in a socket made from steel plates bolted to a wooden arm. The inside of the top plate should be beveled as shown, so it doesn't score the ball. The locking "socket" that the ball fits into is a short piece of 2-in.-dia. pipe, beveled and capped with a disc. A twist on the screw handle will lock this thing up tighter than Dick's hatband.

—*John Stockard, Milledgeville, Ga.*

Bench screws for carving

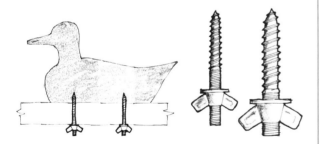

Except for their expense, bench screws are the ideal solution for fixing carving blanks to the workbench. Fortunately, you can duplicate their function for mere pennies. Ask at your local hardware store for hanger bolts. These bolts, which are available in a variety of smaller sizes and lengths, have a wood screw on one end and a machine screw on the other. Replace the nuts that come with the bolts with wing nuts and you have virtually duplicated a $20 bench screw.

The hanger bolts will be more than adequate for holding smaller carvings. But larger carving blanks call for a heftier bench screw. To make one, hacksaw the head off a lag screw as big as you need and thread the shank with a standard die. Fit the screw with washers and a wing nut and you have a monster bench screw. —*Ford Green, San Antonio, Tex.*

Woodcarver's vise

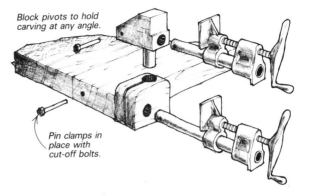

Block pivots to hold carving at any angle.

Pin clamps in place with cut-off bolts.

The pipe-clamp vise shown above makes my woodcarver's clamping system ("Methods of Work," *FWW* #55) even more versatile. Work can be clamped in virtually any position in the top vise, which pivots around 360°. The bottom clamp locks the top vise at the desired angle.

—*Wallace C. Auger, Fairfield, Conn.*

Plastic film protects workbench

If you're tired of cleaning dried glue residue off your workbench, try covering it with a piece of clear plastic film. I use the 4-mil-thick film that is available at many hardware stores for covering windows. Glue drops that fall onto the film dry quickly and, once dry, can be easily cleaned off by pulling the plastic over the edge of your saw table. The residue will peel right off, leaving the film clean for your next project .

—*Marilyn Warrington, Shiloh, Ohio*

Detachable vise pads

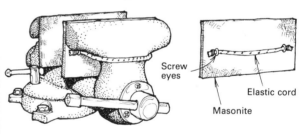

Screw eyes

Elastic cord

Masonite

These easy-to-make jaw pads enable me to adapt my machinist's vise for woodworking, and just as quickly, to switch it back to metalworking. To make the pads, cut two 1/4-in.-thick Masonite or hardwood blocks as wide as the vise jaws and as tall as the distance from the top of the jaws to the vise screw housing. Thread a short length of 3/16-in. elastic cord through two screw eyes on the back of the pads to hold them to the jaws. Secure the ends of the elastic cord by wrapping them with 20-gauge steel wire.

—*George A. Ferrell, Huntsville, Ala.*

Woodcarver's clamping system

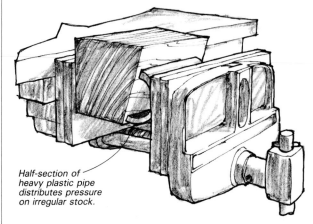

Screw carving blank to anvil.

Insert pipe clamp through swivel block.

Handscrew can be used instead.

Cut-off bolt pins pipe clamp.

Pipe clamp can grip work on its own or with auxilliary devices shown.

I originally designed this clamping system for holding half-size duck carving blanks. With a couple of additions, the system is quite versatile and can be used for many other woodworking jobs as well. The basis is a standard pipe-clamp head mounted on a stubby pipe, about 8 in. long. Drill a pipe-sized hole into the edge of your bench near the corner. If your benchtop is not thick enough to provide a strong lip above the hole, glue a block to the underside of the top to make the total thickness 2 in. or so. Now drill a ½-in. hole from the edge of the benchtop through the pipe and install a sawed-off ½-in. bolt to pin the pipe.

The clamp will serve quite well alone or with a bench dog as a light-duty vise. But two easy-to-build additions increase its uses. One addition, shown in the sketch, is a swiveling block and anvil for carving in the round. Insert the pipe clamp through the hole in the swiveling block before pinning the clamp into the bench. Then, work mounted on the anvil can be turned and swiveled to virtually any angle before the pipe clamp is tightened to lock it in place.

The second addition is simply a standard handscrew drilled so it can be slipped over the pipe. The clamping system can be set up or removed from the bench in just seconds.

—Wallace C. Auger, Fairfield, Conn.

Clamping odd shapes

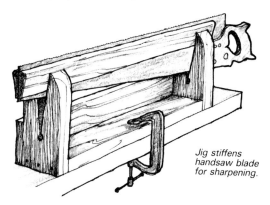

Half-section of heavy plastic pipe distributes pressure on irregular stock.

To hold odd-shaped workpieces in your bench vise, cut a section of heavy plastic or iron pipe in half and place the half-round against the workpiece. The pipe will distribute the pressure and hold the workpiece securely.

—Albert T. Pippi, Baltimore, Md.

Simple saw vise

Sometimes, far from home, your handsaw suddenly encounters a hidden nail. Do you haul the saw back to your shop to resharpen, or do you apply an extra few pounds of elbow grease? Neither solution is very satisfactory. Instead, why not fashion a simple saw vise from a few scraps and resharpen your saw right on the job? All you need are three pieces of 2x4 and a couple of lengths of hardwood 1x2s. There are no critical dimensions, so just use what you have.

Jig stiffens handsaw blade for sharpening.

For most handsaws, an 18-in. long jig is about right. The height of the uprights depends on where you plan to set up the jig; I aim for the sawteeth to be at elbow height for comfortable sharpening. Cut long, matching V-notches in the top of each upright, as shown in the sketch, and drill a hole at the point of the V to reduce the chance of splitting.

Now bevel two 22-in. hardwood 1x2s so that, sandwiched together, they match the angle of the V-notch.

To use the jig, hold the wedge strips on either side of the blade about ½ in. down from the teeth, and tap the blade and strips down into the notches. A tap from below will release the whole thing. —Jim Koch, Stamford, Conn.

Bench vise improved

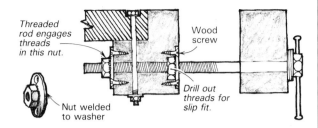

Threaded rod engages threads in this nut.

Wood screw

Drill out threads for slip fit.

Nut welded to washer

Joe Laverti's homemade bench vise *(FWW #37, p. 24)* is a fine idea. But because the heavy steel screws project from the bench, the vise is a potential leg-bruiser. From my school days, I remember a shop teacher hurrying down the aisle between the benches and smacking his leg into an open vise. He was badly injured and the memory has never left me. With a couple of modifications, as shown in the sketch, Laverti's vise can close up like a regular vise and thus be safer.

My vise uses two threaded rods. At the front end of these, I welded a nut and drilled through it to install peened-over bars for handles.   —*Al Glantz, Winthrop, Wash.*

Pipe-clamp bench vise

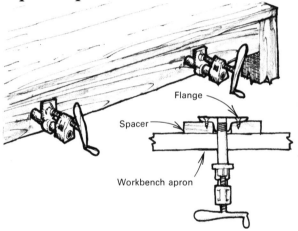

Flange

Spacer

Workbench apron

This simple but effective clamping system can be well adapted to a workbench with a sturdy frame member that runs across the front. The system uses two pipe-clamp heads, two 8-in. pipe nipples and two pipe flanges. The pipe flanges are screwed to the back side of the frame, as shown above, with a plywood spacer between the flange and the frame. The spacer allows the pipe holes through the frame to be the same size as the pipe rather than the larger size that would be required to accommodate the flange. I locate the pipe holes so the clamp heads do not stand above the benchtop, but this decision is based solely on personal preference.   —*Don Rosati, Easton, Conn.*

Pipe-clamp bench slave

I made this bench slave from a pipe clamp, a pipe flange and a wooden base. It is especially useful in holding up the other end of a long board when hand planing in your bench vise. I have seen many wooden versions before, some with notches at various heights, others with lines of pegs and holes to allow a number of height adjustments. Undoubtedly, most of these have been more graceful and handsome than mine, especially when they have been well finished. The prime advantage of my stand is that unlike its wooden cousins, it is infinitely adjustable.

—*Louis Sass, Eola, Ill.*

Threaded-dowel workbench helpers

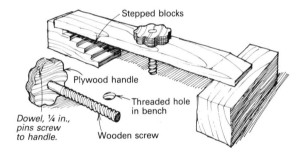

Stepped blocks

Plywood handle

Threaded hole in bench

Dowel, ¼ in., pins screw to handle.

Wooden screw

Here's a way to make that fancy wooden thread-cutting tool earn its keep. First make up a plywood handle in the shape of an oversize faucet handle. Drill and tap the handle, then insert a length of threaded dowel and secure it in the handle with glue and pins. You now have a strong wooden screw that can be adapted to a multitude of clamping, hold-down and fastening jobs. In fact, I've found the wooden screws to be so useful that I have worked out a standardized system using holes tapped into the top of my workbench for fastening jigs and fixtures. The screws are not only strong and adaptable, but they are also ridiculously cheap.   —*Thomas C. Turner, St. John's, Nfld.*

Handtools

Chapter 2

Marking-gauge locking device

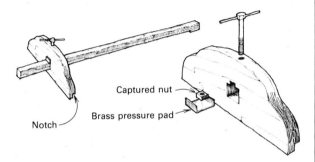

Captured nut

Brass pressure pad

Notch

Many of us make marking gauges and other tools that require a beam to be locked where it slides through the fence. A wedge can be used, but a screw is more positive and more accurate. Of course, screws with wooden threads are nice, but the means for cutting them are uncommon in the average tool kit. So here is an alternative. The version shown in the sketch is a panel gauge (as used by Frank Klausz in *FWW* #70, p. 74), with its fence notched to ride the edges of large panels. This raises the beam above the surface of the panel, cutting down friction and increasing accuracy.

For the screw, you need nothing more than an ordinary ⅜-in. bolt with a square nut. Cut a slot above the beam mortise into which the square nut will slide and be captured. Also enlarge the mortise to allow enough clearance for a pressure pad bent up from ¹⁄₁₆-in.-thick brass. Bend up the ends of the pressure pad high enough to hide the ends of the nut slot. To complete the gauge, drill a hole for the bolt down from the top of the stock into the nut-capturing slot. You may wish to install a ³⁄₁₆-in. rod through the head of the bolt so it can be tightened without a wrench.

—Percy W. Blandford, Stratford-upon-Avon, England

Fixed-position marking gauges

Using fixed-position marking gauges saves me time and tedium while laying out. Here's how I make them: Cut several short pieces from a hacksaw blade. Drill a hole through each blade section and bevel and sharpen one end to a knife edge.

The body of the gauge is a 3-in. by 5-in. block with a mortise in its center. Cut stub tenons on the ends of two pieces of ¾-in. by 1-in. hardwood so they each can be glued into the mortise

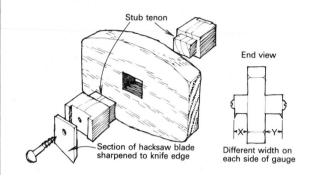

Stub tenon

End view

Section of hacksaw blade
sharpened to knife edge

Different width on
each side of gauge

from opposite sides of the body. Cut each of these two pieces to length so that when a section of hacksaw blade is screwed into a notch in the end, the knife edge will be the desired distance from the body of the gauge. Mount the blade with the bevel facing the block. This way the blade will pull the block into the work and stay on track. When finished, mark the gauge's measurements so you can quickly find the size you're looking for. *—Dennis R. Mitton, Gig Harbour, Wash.*

Shopmade chisel-nose plane

The Stanley #97 chisel-nose plane, originally made for trimming and fitting piano parts, is also useful for trimming off plugs and planing into corners. Its iron is mounted on the front of the plane at a very low 16°

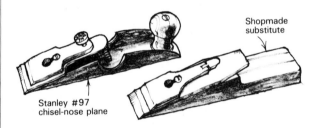

Shopmade
substitute

Stanley #97
chisel-nose plane

Unfortunately, the Stanley #97 is hard to find and collectors often shell out $300 for them. The alternative is to make your own from a plane iron, lever cap, T-nut, ¼-in.-dia. round-head machine screw and a piece of dense 2-in.-thick hardwood, 3 in. wide by 10 in. long. Don't use a regular bench-plane iron, as it's not heavy enough; I used an inlaid tapered iron from an old wooden jointer plane. These heavy old irons are fairly common at flea markets and antique tool sales. Shape the wooden blank, align the screw holes in each and mark the location of the hole. Bore a through hole on the mark perpendicular to the blank's angled face. Enlarge the hole on the plane's sole and install a T-nut. Add a knob if you wish.

—Philip Whitby, Englewood, Colo.

Mortising the throat in wooden planes

Although I started woodworking in the usual way—feeding wood to sophisticated machines, sanding, polishing and spraying—after a few years I was really fed up with the dust and dullness of it. Around that time, I chanced on Krenov's *A Cabinetmaker's Notebook*. Needless to say, the part on planing caught on with me. From then on the stroke of the plane has been the finish of my work.

I like to make my planes from one chunk of wood. It is an unfounded habit of mine, something akin to an instinct, not to make a glueline where it can be avoided. It is certainly quicker to slice the plane body into three sections, cut the cavity in the middle section and glue everything together again. I have done it but do not like it. Instead I use a horizontal mortiser and the following method to cut the throat into my planes.

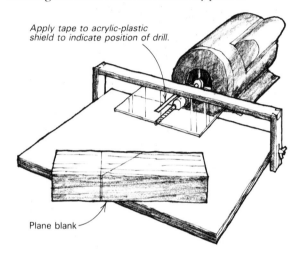

Apply tape to acrylic-plastic shield to indicate position of drill.

Plane blank

First I draw the lines showing the cavity location on all four sides of the plane blank. I remove the bulk of the material from the cavity with a ⅜-in. drill. For this purpose I have made an acrylic-plastic shield that I fit to my shop-made horizontal mortiser. The shield sits parallel to the table just ⅛ in. or so above the plane body. I apply tape to the plastic, directly above the drill, which allows me to "see" the position of the drill within the wood. By aligning the taped outline with the pencil marks on the plane body, I can drill the required holes quite accurately and quickly.

To make the plane's throat, I first drill around the edges of the cavity. Then I waste out the rest, always taking care not to drill too deep. I open the mouth of the plane carefully from the sole with a ⅛-in. drill. Then I clean up the cavity with a chisel and level the plane-iron bed with a float.

—*Stefan During, Texel, Holland*

Ersatz router plane

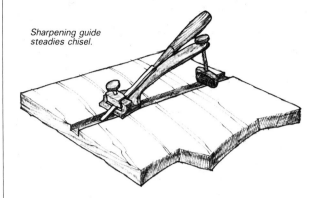

Sharpening guide steadies chisel.

If you don't own a router plane but need to clean out a few dadoes fresh from the tablesaw, try mounting a ¼-in. mortising chisel in a sharpening guide. Adjust the chisel for the desired depth of cut and proceed. Go gently. If you push the tool too hard or too fast the chisel may chatter out of adjustment or make an unwanted submarine dive into the dado.

—*Richard Melloh, Plainfield, N.H.*

Depth stop for backsaw

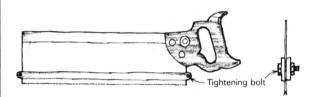

Tightening bolt

This adjustable depth stop for the backsaw aids in cutting accurate dadoes, rabbets, and half-lap joints. The idea is adapted from an antique saw I have. The stop is a couple of lengths of ⅜-in.-thick steel bar stock fitted with bolts on each end to tighten the stop on the blade at the desired setting. Alternatively, I suspect that the two bars could be made of wood if they were crowned slightly in the middle to clamp the entire blade length when tightened. If I have only a few dadoes to make I nearly always use this saw. It is easier and quicker than setting up the tablesaw with dado blades.

—*Bert Whitchurch, Hemet, Calif.*

Milk-jug chisel protectors

A good, fast way to make chisel and knife-edge protectors is with plastic one-gallon milk jugs. Because these jugs are made of a thermoplastic, they can be easily shaped with heat and pressure. I use a propane torch to heat the plastic until it turns clear. Then I press the tool edge straight into the plastic and hold it for a few seconds until the plastic cools. To finish I remove the tool and cut the plastic to the length required. This technique is also good for taking impressions of almost any small object to make plaster casts.

—*Robert Kelton, Saranac Lake, N.Y.*

Preset mortise gauge

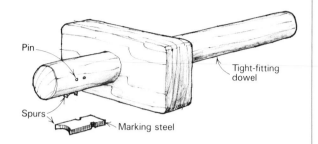

Pin

Tight-fitting dowel

Spurs

Marking steel

I specialize in making chairs, and realized after a few years that my mortise-and-tenon joints were all just about the same thickness, ⅜ in. To save the time spent setting up my adjustable mortise gauge, I made one with a fixed ⅜-in. setting. The most important component is a piece of 1/16-in.-thick steel ground exactly ⅜ in wide and shaped as shown, which provides two marking spurs. The steel should always be sharpened on the inside hollow so the outside dimension is not altered.

Fasten the marking steel into a slot in the dowel with rivets or a small wedge. The dowel should be a hard, stiff piece of wood such as oak. I made the body of the gauge from hornbeam, a dense but non-brittle wood. I omitted the usual wedge for locking the gauge at its setting because the dowel fit so tightly into the body. To adjust the gauge, I tap the dowel with a small hammer. —*Stefan During, Texel, Holland*

Shopmade pull saw

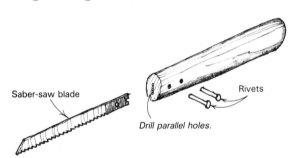

Saber-saw blade

Rivets

Drill parallel holes.

I don't own a saber saw, but I do buy the blades—they make the handiest small saws in my shop.

First, choose a drill bit the same thickness as the sawblade and drill four or five holes side by side in the end of the handle blank. Using the blade as a template, mark the location of the rivet holes on the side of the blank. Now clamp the blade upright in a vise and tap the handle over it. Drill holes where marked, rivet the blade securely in place, and shape the handle to suit.
—*Stefan During, Texel, Holland*

Double scratch stock

A scratch stock is a simple but effective tool for cutting molding patterns on odd-shaped workpieces. But when you scrape against the grain, as is often necessary on curved members such as tripod-table feet, the tool chatters and can roughen surfaces. This double scratch stock solves the problem.

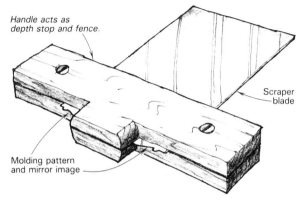

Handle acts as depth stop and fence.

Scraper blade

Molding pattern and mirror image

Grind or file mirror images of the desired pattern on a cabinet scraper, as shown. File the edge to 45°, turn the burr, then mount the blade in a hardwood block that will act as depth stop and fence. Adjust the scraper blade for the proper depth of cut and tighten the screws to lock the blade in place. When you run into contrary grain, simply switch to the other side.
—*John S. Pratt, Avondale Estates, Ga.*

Chisel sheaths from old glove finger

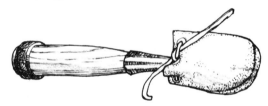

To prevent my chisels, knives and auger bits from damaging each other, I use the thumbs and fingers cut from old pairs of leather work gloves. I punch holes around the opening, and then add eyelets and a length of leather thong to tie the protector on the tool. I'm told that some leathers contain acids that encourage rust, so check your tools once in a while if you plan to try this method for long-term storage. With my everyday tools I've had no problems.
—*Craig S. Walters, Forest Ranch, Calif.*

Making dowels the colonial way

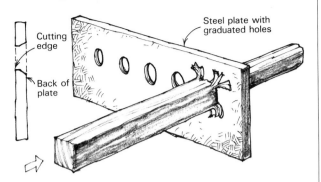

In colonial times wooden pegs were often made by driving a square stick through a round hole in a steel plate. I haven't read much about the technique lately, but it still works. Drill a hole of the desired dowel diameter in a ⅛-in.-thick steel plate, then drill a series of two or three more holes of slightly larger diameters. Countersink the back of each hole, as shown in the sketch, then stone the front surface flat and smooth to provide a good cutting edge.

To use, cut a square stick slightly larger than the final dowel diameter, whittle a tapered point on one end and drive it (from the flat side) through successively smaller holes. In sizes smaller than ⅛ in., it's best to pull the dowel through.

Mild steel is quite satisfactory for this purpose—when the die becomes dull, just drill a new series of holes.

—*H. Norman Capen, Granada Hills, Calif.*

Removing paint-filled screws

I have been a carpenter all my working life and have covered just about all aspects of the trade. A simple but effective trick that's not often seen is a screwdriver modification for removing screws whose slots have been filled in with paint. File a small V right in the tip of the blade. Now just hammer the screwdriver into the slot and unscrew. The V will allow the tip to penetrate the paint, and it doesn't affect the screw-turning aspects at all.

—*Reg Fuller, Turramurra, New South Wales, Australia*

Cutting felt circles

Recently, I made a jewelry chest containing a number of trays with compartments created by drilling spaced holes with a 2⅛-in. Forstner bit. This left me with the problem of cutting numerous 2⅛-in. circles from sheets of self-adhesive felt for lining the cavities. I solved this problem by replacing the pencil in my 8-in. bow compass with a standard X-Acto knife. With this setup, I was able to cut the felt circles with ease in a few minutes.

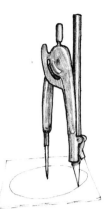

—*Douglas B. Hammer, Solon Springs, Wis.*

Super rasp

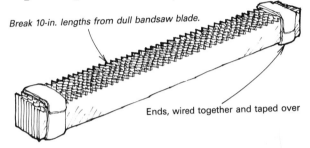

My pack-rat instincts paid off one day when I needed a heavy-duty rasp for a sculpture. Retrieving a broken 1-in.-wide bandsaw blade from the junk pile, I snapped off 10-in. lengths in the vise until I had a 1-in.-thick stack. I bound the ends together with wire and taped over the wire with duct tape. In about 10 minutes, I made the fastest-cutting, easiest-cleaning rasp I've ever had my hands on.

With a little more experimentation, I found that wider blades worked better and staggered teeth made a smoother cut. Blades can be added or subtracted to make rasps of specific widths. To release chips, flex the blades in the middle.

If you make a super rasp, please be careful. This monster eats knuckles with the same appetite that it eats wood.

—*Greg Connell, Lake Elsinore, Calif.*

Canned lubricant

Lightly oiling a handsaw's blade or a plane's sole makes the tool easier to use by reducing friction. Just tightly roll up a 2-in.-wide band of upholsterers' hessian webbing, tuck it into a tuna-fish can and soak with thin machine oil. Then wipe the tool over it, or it over the tool. Resoak the block if it dries out..

—*H.G. Durbin, Porthcawl, England*

Sharpening and Grinding

Chapter 3

Grinding bowl-turning gouges

Gouge · Jig · Setscrew

V-cut provides cam action.

Pull lightly on gouge and pivot counterclockwise.

Borrowing some ideas from fellow turners, I devised this jig for grinding the fingernail shape on bowl or spindle gouges. I find this too tricky to do freehand; you not only have to rotate the gouge while grinding it, but have to move it forward and back.

Start with a 2-in.-square hardwood block, 6 in. long, and turn a 2-in.-dia. cylinder in the middle of the block, leaving the back 2 in. and the front ½ in. of the block square. The rear square is fastened to the base, the front steadies the turning on the bandsaw and will be cut off. Bore a hole along the center of the block equal to the diameter of the gouge to be ground.

Make a V-cut in the cylinder; this provides a forward cam action during grinding to produce the fingernail shape. The depth of the V-cut should be half the diameter of the gouge. To make the cut accurately, scribe two circles around the cylinder half the diameter of the gouge apart. Draw a centerline down the cylinder's length, bisecting the two circles. In plan view, start the cut where the circle nearest you meets the cylinder's surface and end the cut where the farthest circle intersects the centerline. Without pivoting the cylinder, repeat the procedure on the next cut.

The back of the jig is bolted to the base so it can pivot away from the grinding wheel. Drill a small hole for a setscrew through the side of the front cylinder to hold the gouge in place as it's ground.

To use the jig, hold the two cylinders together and push the gouge, flute side up, through both parts and tighten down the setscrew. Adjust the jig so the gouge hits the wheel to produce a 30° bevel angle and clamp the jig down. Lower the gouge into the wheel and rotate it counterclockwise while pulling back on it enough so the two parts of the cylinder stay together.

—*Clif Sessions, Bartow, Fla.*

Sharpening jointer knives

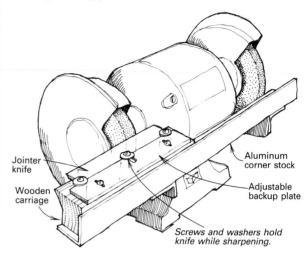

Jointer knife · Wooden carriage · Aluminum corner stock · Adjustable backup plate

Screws and washers hold knife while sharpening.

Although I tried several of the jointer-knife sharpening methods published in this column, none worked well for me. One day, while looking at my bench grinder, I was hit with the idea that if the two tool rests were connected, they would provide an ideal support for a sliding sharpening carriage, even if only one abrasive wheel was used for the actual sharpening. To implement my idea, I spanned both tool rests with a length of 1-in. by 2-in. aluminum corner stock, letting one end extend beyond one of the rests. I fastened the corner stock to the tool rests with flat-head machine screws and wing nuts. Then, I milled a sharpening carriage from hardwood to the profile shown in the sketch so it would fit around the wheel's arc. On top of the carriage there's an adjustable aluminum backup plate and three screws and washers to hold the knife to be sharpened. The carriage's dimensions are such that the blade bevel is ground at a 30° angle.

To use the carriage, install the first knife with the backup plate adjusted so the knife just touches the wheel. Slide the carriage past the wheel with constant light pressure directed downward and backward toward the aluminum guide. For the first few passes, the downward pressure should be minimal. For successive passes, I move the blade toward the wheel by shimming with one, then two and eventually three strips of writing paper between the vertical side of the aluminum guide and the back of the wooden carriage. You might say this is crude, but it works just fine.

—*Henry R. Jaeckel, P.E., Nevada City, Calif.*

Blade covers

After spending several hours honing a new 3-in.-wide fishtail gouge, I considered fabricating a special blade cover to protect the cutting edge. A better idea came to me as I paged through a mail-order supply catalog and spied a product called Plastic Dip, normally used for coating plier handles.

Here's how I made a blade cover with Plastic Dip. First, I spread some Vaseline on the blade so the plastic wouldn't stick to the metal. Then, I dipped the blade in the plastic once a day for four days. After four applications, I trimmed the top of the dip and pulled at it gently. My newly made blade cover came off with a thwock! The dip had molded itself perfectly to the blade, and the Vaseline allowed easy removal. I couldn't ask for a tougher material. Plastic Dip is sold by Leichtung (4944 Commerce Parkway, Cleveland, OH 44128) for about $8 per can.

—*Carl Hungness, Speedway, Ind.*

Improved sharpening-stone box

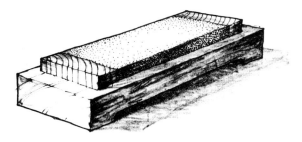

By mortising two endgrain blocks into your sharpening-stone box (one on each end of the stone), you'll provide support for longer sharpening strokes-the tool won't skip abruptly off the edge of the stone at the end of a stroke, and you won't risk catching the edge on the end of the stone at the beginning of a stroke. Longer strokes will reduce wear in the middle of the stone and greatly increase the stone's life.

—*James Gauntlett, Boise, Idaho*

Adapting an in-place knife grinder

Past issues of *FWW* have presented a number of methods for grinding jointer knives. I've never been satisfied with the in-place knife-grinding attachment from my Rockwell 13-in. planer, so I designed a steel base to adapt the in-place grinder for use off of the planer. This fixture provides excellent and professional results. I can also use it to hone my planer knives and grind my joiner knives.

The fixture consists of the knife grinder, a welded steel base and an aluminum knife holder machined to hold the knives at a 36° grinding angle. I bolted my equipment to the steel base by drilling and tapping holes; this is the same way it's attached

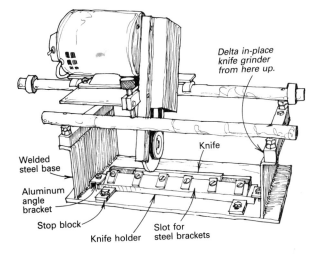

Delta in-place knife grinder from here up.

Knife

Welded steel base

Aluminum angle bracket

Stop block

Knife holder

Slot for steel brackets

to the planer. The knife holder is clamped to the bottom plate of the base with aluminum angle brackets at each end. With steel brackets, clamp the knife to be sharpened to the knife holder. Tighten it in place with bolts threaded into holes drilled and tapped in the angled face of the holder. Adjustable aluminum stop blocks align the knife holder so the knife is centered under the grinding wheel and is parallel to the wheel's travel. The grinder's guide bars are then adjusted so the grinding wheel travels parallel to the knife holder. After making these adjust-

ments, push the stop blocks against the knife holder and bolt them in place. This allows the knife holder to be removed and replaced in the same position for grinding the remaining knives.

—*Earl M. Wintermoyer, Niceville, Fla.*

Foot-powered hand sharpening

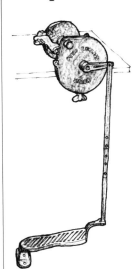

A few weeks back, I walked into a friend's shop and found his new hand grinder in disuse. His excuse was that he just couldn't crank the wheel with one hand and move the tool accurately enough with the other hand to achieve the sensitive, complex grinding required to shape a carving gouge, for example.

Ten minutes later, I'd tied a cord around the handle and to a 2-ft.-long board under my foot to produce a foot crank. Then, not more than a week later, I saw (in the San Joaquin Fine Woodworkers Association's newsletter) a reproduction of a turn-of-the-century advertisement featuring a foot-operated grinder. Bingo—woodworkers were intelligent once! The old grinder featured a hinged treadle and a steel connecting rod that would certainly have worked more smoothly than my cord-and-board crank.

—*Del Stubbs, Chico, Calif.*

Recipe for razor-sharp carving tools

During my 50 years of carving I have collected some 280 edge tools which, especially for the kind of carving I do, must be kept sharp enough to shave with. To prepare the edge, I use three grades of progressively finer India stones. But the real trick is to strop the edge to a mirror finish. For this you will need a couple of pieces of sole leather from the local shoe shop and an abrasive product called Cloverleaf Abrasive Compound, which was originally manufactured for grinding engine valves on Model T Fords. It is a smooth-cutting abrasive suspended in a Vaseline-like jelly. Cloverleaf is still manufactured today in seven different abrasive grades and can be bought in most auto-supply stores. You will need two grades—I use one up from finest and two down from coarsest.

First soak the pieces of sole leather in light lubricating oil. Then rub about a teaspoon of the finer abrasive into the smooth side of one piece and a like amount of the coarser abrasive into the rough side of the other piece. Bend the leather into the profile of the cutting edge and strop both the inside and outside of the carving gouge to produce an incredibly sharp edge.

—*Ford Green, San Antonio, Tex.*

Truing muslin polishing wheels

To true the perimeter of stitched-muslin polishing wheels after they get raggedy and misshapen, chuck a Surform sanding disc in your electric drill. With both the sanding disc and the wheel running, carefully bring the disc into contact with the wheel. The Surform's rotation should be opposite to that of the muslin wheel. Apply the Surform in short bursts so that it doesn't overheat.

—*W. H. Fowler, Anchorage, Alaska*

Perfect edges on rust-pitted tools

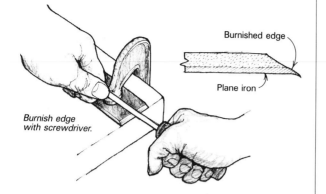

Burnish edge with screwdriver.

Burnished edge

Plane iron

I discovered this sharpening technique while working on an antique laminated-steel plane iron with a rust-pitted back. There just wasn't enough steel to chase those nasty pits to the core to obtain a flawless edge. This technique, which I now use on all my edge tools, burnishes the edge down to provide enough metal for a perfect, work-hardened edge while leaving the back of the tool in its original shape.

Before starting the edge procedure, I gently remove the rust from the old tool with emery cloth, a wire brush or green pot scrubbers. Then, I rough-grind the tool to the proper bevel angle using a hand-cranked grinder and a common silicon-carbide wheel. Next, with the tool clamped to the bench as shown in the sketch, I burnish the edge with a screwdriver shank to produce a curl about 1mm tall. This burnishing operation isn't delicate; rather, it's a rough procedure requiring great pressure, determination, a tightly clamped workpiece and a safely dull edge. Burnish from the corners in to the center to avoid corner breaks. Western tool steel is tenacious, but it'll move if you press hard enough.

Next, I remove the burr and flatten the back of the edge by drawing the tool's back obliquely across a progression of flat stones. Since removing the burnished edge wears stones quickly, you may wish instead to use a sheet of emery cloth oiled or wetted to a sheet of glass. When the back of the edge is flat, you're ready to proceed with honing the edge in normal fashion. —*Generik Tooles, Madison, Wis.*

Honing carriage

Wet-or-dry sandpaper

This simple honing carriage will enable you to hone your plane iron or chisels on a sheet of commonly available wet-or-dry abrasive paper. Select a 2-in.-wide, ½-in.-thick piece of hardwood about 9 in. long. Using a bevel angle of 20°, cut the board in two about 3 in. from one end. Rejoin the pieces with glue and two wood screws into a dogleg, as shown, and attach wheels—mine are nylon pulleys from an old drapery rod.

To use the carriage, screw the plane iron to the dogleg through the slot in the plane iron. The angle of honing can be adjusted from 20° to 30° by raising or lowering the plane iron. Make sure the blade is square to the working surface so that the entire bevel is honed evenly. Place a sheet of waterproof 240-grit silicon-carbide abrasive paper on a hard flat surface such as a sheet of glass or a piece of Formica. Keep the sandpaper flushed with water while you move the honing guide forward and back with the tool's bevel resting on the wet abrasive paper. Soon the sharpening action will produce a wire edge. Remove the blade from the carriage and strop by hand.

—*Tom Froblich, North Miami Beach, Fla.*

Sharpening center punches

Here's how to put a perfect cone-shaped point on the end of a punch. You'll need a grinder fitted with a 6-in. coarse wheel, a small electric drill and a small piece of heavy cardboard. Wet the cardboard thoroughly, wrap it around the barrel of the punch and let it dry overnight in place. The next morning spring open the cardboard and grease the inside lightly. This cardboard tube is your guide bearing and friction insulation. Now chuck the punch in the drill and, holding the spinning punch with the greased cardboard tube, bring the punch to the rotating grinding wheel. Grind the point to shape on the edge of the wheel but finish on the side.

—*Ford Green, San Antonio, Tex.*

Low-cost grinder misting system

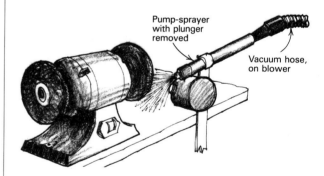

Pump-sprayer with plunger removed

Vacuum hose, on blower

Charles Riordan describes a simple, effective sharpening system (*FWW*#39) that uses a compressed-air-powered misting device to cool the grind. Having neither the money nor the need for an air compressor in my shop, I improvised the misting system shown here. Start with a hand-powered sprayer, known in many locales as a "flit gun," and remove the plunger. Put your shop vac's hose on the blower side and insert the hose's nozzle in the pump cylinder. Now simply fill the flit-gun canister with water, mount the device on a stand near your grinder and turn on the vacuum. Voilà! Low-cost, low-tech misting.

—*Peter S. Birnbaum, Sebastopol, Calif.*

Jig for honing two jointer knives

My Inca jointer has two 10¼-in. knives, which cost $12 to sharpen. Not satisfied with the price and inconvenience of that, or with the homemade sharpening devices I'd seen in the "Methods of Work" column, I built a device that allows manual sharpening of both blades at once. This mostly-wooden jig, shown in the sketch below, is inexpensive, and accurate if smooth strokes are used along the full length of the knives.

To make the jig, start with a piece of straight-grained, 2-in. square hardwood as long as your knives. Chamfer the top of the block so that the bevels will be parallel to the top of

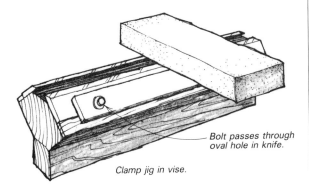

Bolt passes through oval hole in knife.

Clamp jig in vise.

the block. Glue on wooden strips, slightly thinner than the thickness of the blades, to act as stops. Make sure these strips align the blades so that the beveled edges are in the same plane. Cut and drill steel strips and install them as shown in the sketch to hold the blades securely.

For safety's sake wear a glove and be careful. A slip could cause a nasty cut.　　　　　　　　*—John Toffaletti, Durham, N.C.*

Sharpening skew chisels

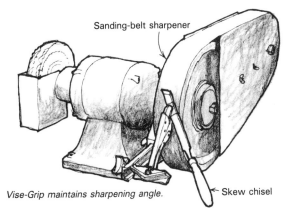

Sanding-belt sharpener

Vise-Grip maintains sharpening angle.　　　← Skew chisel

Some sharpening setups have a special tool rest to support the butt of the tool's handle, which keeps the cutting edge at the proper sharpening angle. The idea works great for straight plane irons and chisels, but presents problems for skewed tools. To put the skewed tool at the proper angle on the sharpening belt, the handle must be pulled to the side and held in midair, unsupported.

To solve the problem, I clamp a pair of Vise-Grips to the tool as shown in the sketch. I protect the chisel blade with a wrap of duct tape. If I have to disconnect the Vise-Grips during the grinding process, the imprint on the tape allows me to place the grips in the exact position again to complete the job.
　　　　　　　　　　　　　　　—Norman Vandal, Roxbury, Vt.

Regrinding chisels on a disc grinder

Here's a method that I think is unbeatable for regrinding chisels and plane irons. I clamp my Makita portable disc grinder in a Workmate vise (with a couple of pine jaws) and use a board, shimming it if necessary, to produce a surface flush with the grinding disc. Then I clamp the chisel in a honing guide (I have a Japanese model with handles, as shown, available from Garrett Wade). The roller of the honing guide runs on the board while the blade is ground by the disc. I keep the blade from overheating by frequently dipping it in water. Since the blade remains in the guide, I can return it to the grinding disc at precisely the same angle.

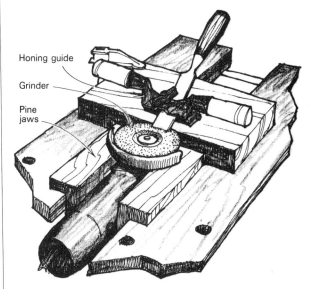

Honing guide

Grinder

Pine jaws

I prefer the Japanese honing guide to others because the spokeshave-like handles allow me to rock the guide slightly from side to side to produce a crowned edge on plane irons. This would be possible with other guides, but the handles on the Japanese guide provide greater control.

With this method, I have reground bevels that are indistinguishable from those ground by the factory. It has changed a frustrating and difficult task into one I can accomplish with precision and ease.
　　　　　　　　　　　—Robert B. Campenot, Freeville, N.Y.

Sharpening system

I've replaced the sharpening stone and strop in my shop with a two-wheel buffer and two abrasive compounds commonly used by knifemakers and gunsmiths. First I grind the tool's edge on a regular grinding wheel, then I buff the edge on a muslin buffing wheel loaded with a greaseless buffing compound manufactured by Lea Manufacturing Co. (available from Badger Shooter's Supply, Box 397, Owen, Wis. 54460; 715-229-2101). Even its fine grade cuts fast enough to send a few sparks flying, so I quench the tool often to prevent heat buildup. Next, I polish the edge with white No. 555 Polish-O-Ray (available from Brownells Inc., Route 2, Box 1, Montezuma, Iowa 50171; 515-623-5401). Alternate polishing the top and bottom of the cutting edge. Only a light touch is required to finish the edge to perfection.　　　　　　*—Robert Mordini, Edmond, Okla.*

Shop Aids

Chapter 4

Disassembling old tabletops

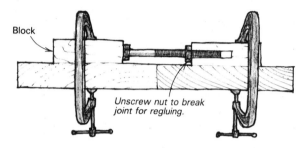

Block

Unscrew nut to break joint for regluing.

This technique is quite effective for disassembling old tabletops for regluing. It requires two scraps of hardwood, one of which is drilled to accept a hex-head machine bolt and nut, as shown in the sketch. When you unscrew the nut, it exerts pressure against the clamped blocks, forcing them, and the glue joint, apart. As pressure builds, place a piece of scrap over the joint and hammer the scrap to jolt the glue joint. This technique puts tremendous declamping pressure in just the right spot without damage to the tabletop.   —*Frank D. Hart, Plainfield, Ind.*

Wall-mounting cabinets

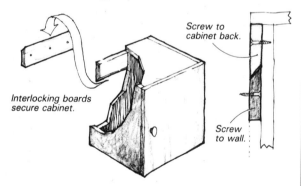

Screw to cabinet back.

Interlocking boards secure cabinet.

Screw to wall.

This simple method for hanging wall cabinets is fast, easy and accurate. To make the mount, rip a ¾-in.-thick board in two at a 45° angle. Screw one half to the wall to form a perch and screw the other half to the cabinet back, which should be recessed ¾ in., as shown. Then just slip the cabinet over the perch board—a one-person operation. As a bonus, the cabinet can be easily removed whenever needed.

—*George C. Muller, Union, N.J.*

Installing small brass knobs

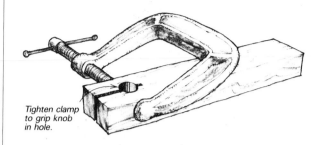

Tighten clamp to grip knob in hole.

Small brass knobs with threaded shanks can be difficult to install without marring their finish, especially in very hard woods. I solved the problem with this grip made from scrapwood. Use a small C-clamp to squeeze the knob in the hole, but take care that the clamp doesn't drag on the wood and scratch it. The wood scrap acts as a non-marring handle, allowing easy installation of the knob.   —*Mac Campbell, Harvey Station, N.B.*

Alignment block for accurate threading

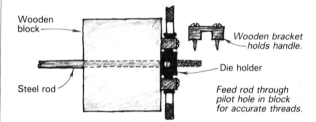

Wooden block

Wooden bracket holds handle.

Die holder

Steel rod

Feed rod through pilot hole in block for accurate threads.

I wanted to make my own maple handscrews, but found that freehand threading of the ⁵⁄₁₆-in. steel rods invariably resulted in erratic thread patterns and wobbly clamps. I solved the problem by drilling a pilot hole through a 2-in. block and fastening my die holder directly over the hole with little maple brackets. To ensure firm clamping pressure, the cutouts in the brackets should be a fraction shallower than the height of the handles. I now get perfect threads every time, both right-hand and left-hand, and have produced a number of beautifully functional clamps at a fraction of the cost of store-bought tools.
—*Chris Clark, Winnipeg, Man.*

Versatile mylar

I recently discovered a drafting "paper" made of transparent mylar, which has proved to have many interesting applications around my woodshop. It is available in rolls or sheets from art-supply stores, and comes in various thicknesses from 0.003 in. to 0.008 in.—this makes it handy as shim stock, and the heavier weight makes good template material, too. One side is frosted so that it will accept pencil or ink, and the other side is glossy. You can buy it clear or with a graph-paper pattern for making drawings. I find that for small projects I can draw the plan on the frosted side, turn the sheet over, and glue up the work on the shiny side—the glue won't stick to it and I can align the pieces perfectly with the drawing. Mylar cuts well with a sharp knife or scissors, is durable and has very good dimensional stability. It is also superior to waxed paper for protecting surfaces from glue squeeze-out because it is stiff, doesn't wrinkle and tends to stay put.

—*Norm Capen, Granada Hills, Calif.*

Homebuilt sawdust-burning shop heater

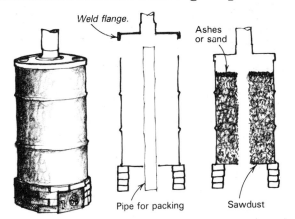

Weld flange.

Ashes or sand

Pipe for packing

Sawdust

If you've ever looked at the huge pile of sawdust and planer shavings that even a small shop produces and wished you could burn it for heat, here is an inexpensive but efficient solution based on an ordinary 55-gal. steel drum.

Construction of the stove is relatively simple. Because you have to open up the stove every day for loading, what you're after is a removable lid that can be easily assembled or disassembled from the stovepipe. First cut the top off the drum right below the lip. Fashion a retainer ring of strap iron and weld it around the top to produce a removable lid, as shown in the sketch. Cut a hole in the center of the lid and attach a 6-in. stovepipe adapter flange to the middle of the top. You will also need to cut a 4-in. hole in the center of the bottom of the drum for the fuel packing pipe, explained below. Set the stove up on an airtight ring of firebricks (use fireclay for mortar) laid right on your cement floor. If possible, add a scrounged door from an old woodstove to regulate air and to provide an access for cleaning out ashes—this was an improvement we added the second year; the first year we blocked off the air with the lid of a 5-gal. paint can.

How this stove works is truly amazing. The secret is leaving a chimney hole right through the fuel after it is packed. To do this you just remove the stove's lid and insert the "fuel packing form" (a length of 3½-in. plastic sewer pipe) down through the hole in the bottom so that it sticks up out of the top of the drum. Now load the drum with planer chips, sawdust, floor sweepings—anything that will burn. Pack it down tightly around the plastic pipe until the drum is full to about 5 in. or 10 in. from the top. Sprinkle the top of the packed sawdust with sand or ashes so it won't burn anywhere but in the middle. Pull the plastic pipe out of the packed fuel, replace the lid and light a small fire underneath the barrel.

A full drum will heat our 3,500-sq.-ft. North Carolina shop for eight hours with no attention. Since we are burning kiln-dried wood chips, the flue gases are clean and combustion is complete. When the stove is going strong, no smoke comes from our chimney—only clear, hot gases.

—*Paul G. Caron, Cashiers, N.C.*

Improved forge design

When I began making my own tools a few years ago, I decided to build a forge using a barbeque grill and a vacuum-cleaner blower, based on a design in *FWW #9*. I had some problems with that forge, but the one shown here, intended for outside use, eliminates them.

The main components are an old barbeque grill pan, firebricks for the firebox, a 55-gallon drum that acts as a hood, and two lengths of 6-in. stove pipe to ensure a good draft.

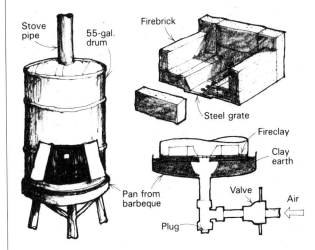

Stove pipe

55-gal. drum

Firebrick

Steel grate

Fireclay

Clay earth

Pan from barbeque

Valve

Air

Plug

Construct the forge as shown in the sketch, connecting the blower with ordinary steel pipe. Install a valve in the blower line (I used a simple sliding wooden plate) to vary or cut off the air, and set a plug at the bottom so you will be able to remove ashes when necessary.

The size of the firepit can be changed, of course, but the configuration illustrated has been the best for my work. Any place that isn't exposed to heat (below the firebrick) can be filled and sealed with hard-packed earth and red clay, of which there is an abundance here in Arkansas. This will save a lot of fireclay, which I used only to seal around the firepit.

After the forge is together, build a roaring fire in the thing until it's good and hot. Check the next day for air leaks by turning on the blower and hanging a loose thread over suspicious cracks. Patch these with clay mud or fireclay.

Although the barrel hood provides enough shade so that I can usually judge forging heat by eye during full daylight, I found it best to save tricky work until evening. The hood also provides good protection from wind.

In working the forge with small pieces of steel, you can save coal by damping the back portion of the fire with water. If the work is longer than the depth of the forge, remove the rear firebricks and push the work through the small rear opening cut into the back side of the barrel.

A couple of safety reminders are to use only firebricks in the firepit—ordinary bricks may explode when heated rapidly. Also, bend the corners of the cut-out door in the barrel and smooth any rough edges. In the flurry of forging, it's easy to run into sharp edges. —*Jim Young, Omaha, Ark.*

PEG vat from scavenged water heater

To make an inexpensive heated unit for impregnating green wood with PEG (polyethylene glycol), I went to a nearby plumbing supplier and scavenged an old electric water heater from their "boneyard." I removed the outer shell and cut the top off the tank, leaving an open, 20-in. deep tank. I placed a metal rack in the tank to support the PEG pail and protect the heating element. I enclosed the heater in an insulated plywood box fitted with large casters.

To use the vat, I place a heavy-duty, PEG-filled rubber garbage pail in the tank, fill the tank with water (like a double boiler) and turn on the thermostat. The heater works great, and my only expense was the garbage pail.

—*Mark Pleune, Suttons Bay, Mich.*

Homemade tank for heating PEG

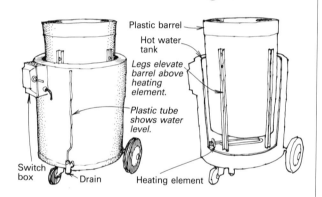

Plastic barrel

Hot water tank

Legs elevate barrel above heating element.

Plastic tube shows water level.

Switch box

Drain

Heating element

I've found soaking green wood in PEG 1000 helps reduce shrinkage of the stock I use for lathe projects. Heating the PEG shortens the time the wood must be soaked, so I built a makeshift double boiler to do the job. The main components are a salvaged electric water heater, a 30-gal. plastic barrel, a couple of lawnmower wheels and an old lever switch box.

The construction is simple. Remove the hot-water tank from its jacket and saw both the tank and the jacket in half. Install the portion of the tank with the heating element and thermostat back into one of the half jackets and pack the space between the two cylinders with fiberglass insulation. Bolt four aluminum legs to the plastic barrel to elevate it above the heating element. Mount the switch box to the outside of the heater and wire it to the thermostat. A drain pipe with a T-shape pipe fitting is screwed into the bottom of the tank. The bottom of the T is plugged and the top is connected to a clear plastic tube that acts as a water-level sight glass.

Now, fill the hot water tank with water and the plastic barrel with PEG 1000. Set the thermostat to keep the PEG solution at the manufacturer's recommended temperature.

—*Charles Manning, Port Townsend, Wash.*

Lumber dolly

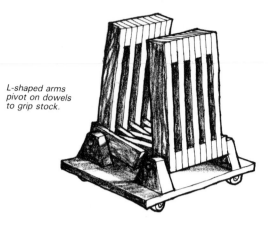

L-shaped arms pivot on dowels to grip stock.

Working single-handed in a small workshop, I found it tiring and awkward to move numbers of 4x8 sheets of plywood or large planks from the delivery truck to my machine area. So I built two of the "bogies" shown here from scraps and inexpensive heavy-duty casters. They have saved me hours of back-breaking work.

First I laminated L-shaped blocks, about 18 in. tall and 8 in. wide at the base, from plywood. Then I drilled a ¾-in. hole through the blocks and mounted them to the base so they could pivot on dowels, as shown.

To use the dolly, spread the arms of the blocks to take the sheet of plywood, which is lowered in. The weight of the wood then levers the arms down to clamp the plywood in place. One dolly, placed in the center, is enough for most loads, but for extra-heavy or awkward pieces you can use two.

—*Chris Yonge, Edinburgh, U.K.*

Plywood keeper

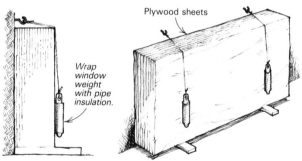

Plywood sheets

Wrap window weight with pipe insulation.

I've used this method on stacks of plywood up to 30 sheets thick. Sink two eyescrews into the wall about 51 in. off the floor. Tie two sash weights to a piece of string and suspend each weight from an eyescrew. Cover the weights with foam pipe insulation to keep them from marring the plywood.

—*John R. Thiesen, Cheektowaga, N.Y.*

Plywood carrying handle

Anyone who has single-handedly maneuvered a full sheet of plywood or drywall through a congested worksite or a doorway will appreciate this easily made gadget. The lifter is nothing more than a foot-long, V-grooved block screwed to a scrap of ¼-in. plywood. Adjust the length of the sash cord so the carrier is a few inches off the ground with your

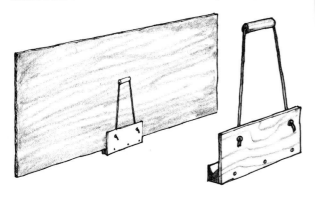

arm fully extended. To use, reach over the plywood sheet to hook the lifter under the lower edge into the center of the sheet. Lift and carry with one hand—the other hand remains free to open doors.

—*G. O. Hoffmann, Cheshire, Conn.*

Two storage containers

Recently Wilson began packaging tennis balls in unbreakable clear plastic containers with removable plastic lids. These containers make excellent storage bins for nails, screws and the like. If you're not a tennis player, just scrounge the trash cans at your nearest tennis court or ask a local club pro to save some for you.

—*R. B. Hurley, Williamsburg, Va.*

Coffee cans with plastic lids are ideal storage containers for nails, etc., except for one problem—you can't tell what's inside. To remedy, fold the lid slightly and poke a nail in one side and out the other (like a safety pin in a baby's diaper) to leave a sample on the lid readily identifying the contents. —*Jeris Chamey, Ponca City, Okla.*

Nail jugs

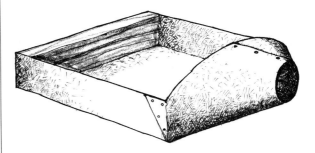

Milk containers hold nails.

When my new house was completed I collected quite a variety of half-full nail bags from the site. To make the collection readily available, I trimmed the necks off plastic milk cartons, marked each jug with the size of its contents and stocked them neatly on wall shelves. The jugs are convenient, durable and ready to transport to any project.

—*Ralph E. Hall, Pisgah Forest, N.C.*

Sorting pan and funnel

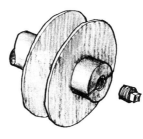

I keep a wide assortment of screws, nuts and bolts in cans with several sizes in each can. To find the item I want, I dump the can's contents into the sorting pan, where it is easy to find what I need, then I use the built-in funnel to pour the contents back into their container.

I used sheet aluminum to make the pan because it is more ductile than galvanized steel. The drawing gives the basic idea, but it has been simplified a little for clarity—a couple of refinements on my pan include cove molding glued around the bottom so things don't lodge in corners, and a stiffened top edge, made by rolling the aluminum around a length of wire.

—*L. Byron Burney, Raleigh, N.C.*

Locking a pulley on its shaft

When all else fails, here's how to lock a pulley to a shaft. First, drill and tap the shaft with ⅛-in. pipe thread. Split the end of the shaft with a sawcut. Replace the pulley on the shaft and screw a tapered pipe plug in the tapped hole to expand the shaft.

—*Douglas M. Ryan, Santa Clara, Calif.*

Cleat-system shop organizer

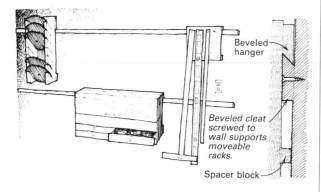

Beveled hanger

Beveled cleat screwed to wall supports moveable racks.

Spacer block

Being a compulsive organizer, I have moved things around in my shop many times and I anticipate more moves in the future. To accommodate all this rearranging, I have come up with a cleat system that makes practically everything in my shop portable. The system consists of two beveled cleats attached to the shop walls. One cleat is attached 40 in. from the floor, a good working height, and the other at 64 in., a good hanging height. Any item I want to attach to the wall is fitted with a reverse-bevel hanger, as shown in the sketch. I use the system to hang my toolbox, router box and drill box, to fasten a grinder to the wall, to position my work lights and to attach hooks for rules and brooms. I use the cleats to hang everything that can be used in several locations. Later, I plan to build an identical cleat system inside a panel truck so I can transfer equipment between shop and truck quickly and neatly.

—*John Loughrey, Madison, Wis.*

Cord-loop storage

Here's a looped-cord storage system that has proved to be a great way to put empty wall space to use. It can be made to work with just about any kind of shop item. I use it to hang tools, bags of hardware or tin cans holding small wood parts. The last time I counted, I had about 50 different things suspended on my shop walls, and they are all plainly visible and easy to get at. I much prefer this to having everything cluttered in corners, hidden in drawers or under my bench. And, an empty cord hanging on its nail tells me that something isn't where it should be

—*Don H. Anderson,
Sequim, Wash.*

Low-cost footswitch

Weld.

Shopmade steel ramp

A foot-operated switch is not only a convenience, but also makes many woodworking operations safer. I was able to make mine at a reasonable cost from hardware that I obtained locally. The main part of the unit is a foot-operated starter switch from a farm tractor (I used International-Harvester part No. 64931-h). It has a 400-amp capacity, more than enough for any 110-volt, single-phase motor found in the wood shop.

Begin construction by brazing or welding a ramp-shaped steel plate to a regular surface-mount electrical box. Drill a hole in the ramp and mount the switch just below the box. Wire a standard 110-volt receptacle into the box using heavy (No. 12 or No. 14) electrical cord and a box connector. Connect the cord's neutral and ground wires to the receptacle but route the "hot" or black wire through the switch. Tape the switch connections well to insulate them.

The switch will operate only when foot pressure is applied and thus functions like a "dead-man" control found on industrial machinery. Locate the switch where no one will step on it accidentally, or if that seems difficult, construct a safety cover. You may wish to fasten a wooden block around the ramp to allow a more comfortable foot angle.

—*Robert L. Koch, Tarkio, Mo.*

Belt tightener

Often a power tool belt will slip just when it's needed most. You can keep the old stretched belt tight (until it can be replaced) by installing a wooden idler pulley similar to the

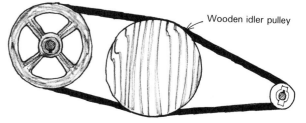

Wooden idler pulley

one shown in the drawing above. Turn the wooden disc on a lathe so its diameter is larger than either pulley on your equipment. Use a skew chisel to cut a V-groove in the edge to fit the profile of the belt. Pull the belt apart slightly and insert the pulley. In operation, the free-running idler pulley will move up and down seeking its own invisible center.

—*Donald F. Kinnaman, Phoenix, Ariz.*

Securing large vacuum bags

Here's my trick for preventing large dust-collector bags from popping off their flanges. I cut a 4-in.-dia. hose clamp into two pieces that I then pop-rivet to the ends of an appropriate length of discarded metal band-strapping. This gives me, in effect, a super-long hose clamp that can be tightened quickly around the joint where bag and machine meet. I use the quick-release-type hose clamp, which makes the device very convenient when removing and replacing the vacuum's bags.

—*James Christo, Jamestown, N.Y.*

Homemade rubber motor mount

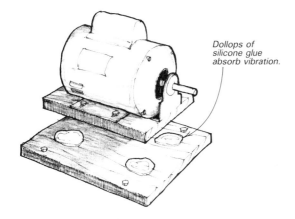

Dollops of
silicone glue
absorb vibration.

A rigidly mounted motor can set up annoying vibrations and cause an entire machine to buzz. Here's how to mount the motor on rubber, which will usually cure any vibration problem. Simply glue two plywood motor-mount boards together with four big dollops of silicone-rubber adhesive. (The rubber seems to adhere better if the boards are varnished first.) To keep the rubber from squeezing out, place ¼-in. spacers between the two boards until the rubber cures. Now, screw the motor to the top board and fasten the bottom board to your machine base. The rubber will insulate the machine from vibration.

—*Bill Webster, Chillicothe, Ill.*

Sharing motors

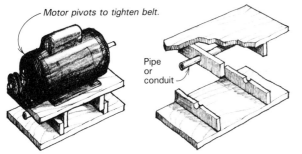

Motor pivots to tighten belt.

Pipe
or
conduit

When I recently purchased some used equipment from the widow of a life-long woodworker, I found it unusual that several of his machines were missing a motor. But, when the motor almost fell off a machine I was carrying, I realized what the woodworker had done. On several of his little-used machines he had installed the same shop-made motor mount shown in the sketch, and consequently he could move a single motor from one tool to the next as needed. The mount is designed so that the motor pivots in the saddle and tensions the belt with its weight.

—*Dan Miller, Elgin, Ill.*

Venturi-box dust catcher

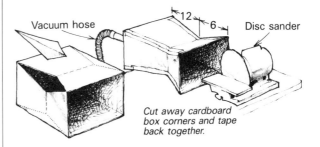

Vacuum hose   12   6   Disc sander

Cut away cardboard
box corners and tape
back together.

My venturi box is an improvement over the standard box-like hoods that are normally used with shop vacuums in dust-collection hookups. I use it to catch dust thrown by my bench-mounted disc sander. Try it at various locations behind the sander wheel until you find the most efficient spot. Air drawn through the box speeds up at the constriction, creating a pressure drop in the rear half of the box, effectively increasing the pull from the shop vacuum. Use any heavy, smooth cardboard box, cut away portions of each corner to produce the double taper shown and reassemble the box with duct tape.

—*Gordon Baxter, Beaumont, Tex.*

Dust collector

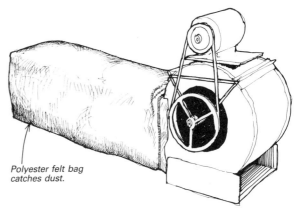

Polyester felt bag
catches dust.

When I went shopping for a shop vacuum system, the culmination of the salesman's pitch concerned the unit's large blower, which sucked the dust out of the entire shop. I realized then that I did not need another vacuum in the shop, simply a dust eliminator. Turning to an article by Mac Campbell in *FWW on The Small Workshop*, I found the key: a bag made from polyester felt. The material, available from most large retail fabric outlets, allows air to flow through, but it catches dust, similar to the way a filter bag works in a vacuum cleaner. I combined a homesewn bag with a discarded squirrel-cage blower, and the rest is history. You would not believe the dust this thing sucks up. I keep the unit on the floor and direct my sweepings toward the inlet; when I'm sanding, I stay close to the blower. All the dust that normally stays in the air for minutes and powders every inch of the shop is sucked up instantly.

—*Thomas C. Turner, St. John's, Nfld.*

Blast gates

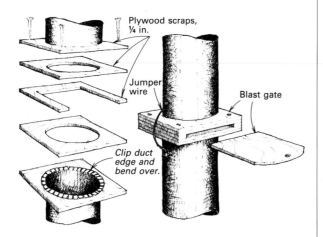

Plywood scraps, ¼ in.

Jumper wire

Blast gate

Clip duct edge and bend over.

After reading about dust-collection systems, I decided to install a system in my shop based on Grizzly's four-bag portable unit. First I fastened a 6-in. main line across the ceiling of my shop with 4-in. branch lines, fitted with blast gates, running down to each of my machines.

I built these blast gates by laminating pieces of ¼-in. plywood scraps as shown in the sketch. To attach the blast gates, I make cuts ¾ in. long, about every ½ in. around the perimeter of the duct, and then bend the tabs over. I slide a piece of plywood up to the bend on each length of duct, then screw through these pieces and the pieces of plywood between them. Although the gates work very well, they break the electrical continuity of the pipes, which can result in a sawdust-igniting buildup of static electricity. I solve this problem by installing a jumper wire from gate to gate.

—*Mike Cole, Coeur d'Alene, Idaho*

Cleaning sawblades, two ways

Sawblades will stay clean longer and clean up faster next time if sprayed with a kitchen non-stick product such as PAM. While spraying, hold a piece of carboard over the blades to catch the mist thrown off, then saw a scrap or two to clear the excess oil from the blade before you use it on good stock.

—*David L. Wiseley, Waters, Mich.*

To clean the gummy buildup on sawblades, spray the blade with oven cleaner. I use the foaming type that contains 4% lye. Let the sprayed blade stand a while until the gummy deposit lifts, then rinse under the tap. Oven cleaner is powerful, caustic stuff, so observe the warnings on the label. Do not use on aluminum tools. —*G. V. Mumford, Ventura, Calif.*

Removing rust with vinegar

I have used this method for removing and preventing rust in my old-tool shop for several years. Disassemble the tool and soak it in full-strength natural vinegar. Store-bought vinegar is standardized at 4% and is not strong enough. Find apple cider with no preservatives, add some "mother" from old vinegar (the cobweb-like stuff) and, in time, the cider will turn to vinegar. After soaking the tool overnight, rinse it under the tap to remove most of the rust. Lightly dress with a wire brush. When dry, spray the tool with transparent aerosol shoe polish to seal the metal and prevent further rusting.

—*Charles W. Whitney, Mt. Liberty, Ohio*

Vacuum screening ramp

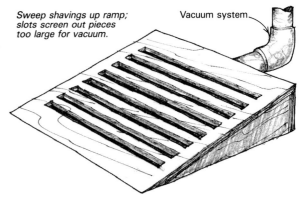

Sweep shavings up ramp; slots screen out pieces too large for vacuum.

Vacuum system

Even in shops with efficient dust-collection systems, there are always piles of sawdust and shavings that must be swept up with a broom. Here's a handy screening ramp to speed up your cleanup. With ½-in. plywood, fabricate a wedge-shaped box with 1-in. slots cut into the top. Attach the ramp to your vacuum system through a hole in the back. Now, simply sweep your piles up the ramp. Any piece too large for your dust collector's digestive system will be filtered out by the slots.

—*Ralph Bell, Ashford, Wash.*

Dust-collection system improvements

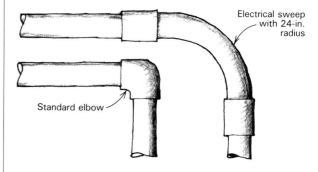

Electrical sweep with 24-in. radius

Standard elbow

When I put together my dust-collection system using common 4-in. PVC pipe I found that, with all the 90° elbows that were necessary, the air flow was restricted and inefficient. Then, at an electrical-supply house, I discovered a special 24-in.-radius PVC elbow called an electrical sweep. The new wide-turn elbows have solved my air flow inefficiencies.

—*John S. Gallis, Deer Park, N.Y.*

Two non-slip push blocks

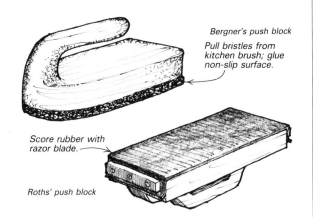

Bergner's push block

Pull bristles from kitchen brush; glue non-slip surface.

Score rubber with razor blade.

Roths' push block

Along with my new jointer I wanted to purchase a set of push blocks—the kind with a molded plastic handle and ¼-in.-thick black foam material on the bottom. But when I found the set was priced at $16, I promptly left the store without it. The next day I happened to bump into a kitchen-brush sale and realized that, except for the bristles on the bottom, the $.55 brushes were virtually identical to the expensive push blocks. So I bought a pair, pulled out the bristles with pliers and glued a Scotch-Brite pad to the bottom for a non-slip surface. Felt and sandpaper or dense foam could have been used equally well. The total project took 30 minutes and cost $2.50 to complete.

—*Mitch Bergner, St. Louis Park, Minn.*

When I needed a non-slip, non-mar push block for pushing panels through the shaper I borrowed an idea from boat shoes. I attached an inner-tube scrap to the bottom of a shopmade block and scored the tube with a razor blade about ¹⁄₃₂ in. deep every ¼ in. or so. The slices open up slightly under pressure and really grab the wood. —*Mike Roths, Vinton, Iowa*

Kitchen baster handy in shop

Transferring lacquer from a gallon can to your sprayer cup needn't be messy and awkward. Use a common kitchen baster like a jumbo eyedropper to transfer the finish. To maintain domestic tranquility, don't steal the baster from your kitchen. Rather, go buy your own, making sure the body is nylon so that it is impervious to lacquer and lacquer thinner. The baster is useful in cleanup also. Use it to squirt solvent through gun orifices. —*Chuck Anderson, Porterville, Calif.*

Measuring and Marking

Chapter 5

Wall-hung right-angle marker

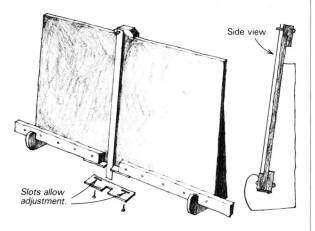

Side view

Slots allow adjustment.

In my modelmaking/prototype shop much of our layout work requires quick, accurate right-angle scribe marks on thin materials ($\frac{1}{16}$ in. to $\frac{1}{2}$ in.). Frustrated with inaccurate and easy-to-knock-out-of-square framing squares, we built this wall-hung right-angle scribing unit that can accommodate materials as wide as 34 in. The main part of the unit is a 3-ft.-tall, 4-ft.-wide panel of cabinet-grade particleboard fitted with a two-piece 1x2 hardwood ledge screwed to the bottom edge. Other components include a stainless-steel ruler that hangs from a pin at the top and is indexed by a notched plate at the bottom.

We turned a threaded rod to make the ruler pin; it must fit the hole in the ruler exactly. At the bottom of the notch-plate assembly is the key to the unit's accuracy, a $\frac{1}{8}$-in.-thick steel plate about 3 in. wide and 6 in. long. File a notch in the plate carefully, so that it is just as wide as the ruler and no more. Screw the notch-plate to the bottom of the assembly through slotted holes so that the ruler notch can be adjusted left and right. You can trial-and-error the ruler into perfect square by scribing a line on a test piece and then flipping the test piece 180°. If the scribed line on the flipped test piece matches the ruler, it is square. If not, adjust the notch plate and try again.

The unit works best when it is tilted back from vertical. The drawing shows a quick, if crude, way to support it with two wall brackets. This allows it to be lifted out if there is a need to use it elsewhere. —*Ed Stringham, East Bethany, N.Y.*

Dividing a circle

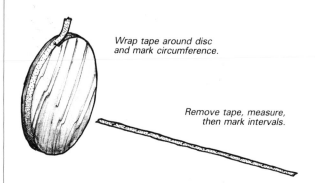

Wrap tape around disc and mark circumference.

Remove tape, measure, then mark intervals.

Here's an accurate way to lay out equally spaced intervals on the circumference of a circle—without the aid of dividers, protractors, indexing heads or geometry skills. Simply wrap masking tape around the circle and mark where the tail of the tape overlaps the starting point. Remove the tape and fasten it to a flat, clean tabletop. Now, measure the distance between the marks to get the circle's circumference and divide this distance by the number of intervals you want. Next, lay out the intervals on the tape, reapply the tape to the workpiece and mark each interval's location on the workpiece circumference by piercing the tape with an awl.
—*Randall Bishop, Christiansburg, Va.*

Laying out equidistant intervals

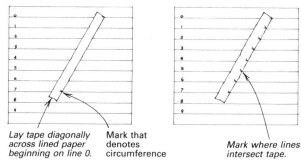

Lay tape diagonally across lined paper beginning on line 0.

Mark that denotes circumference

Mark where lines intersect tape.

Here's an addendum to Randall Bishop's tip for using tape to divide the circumference of a circle into equally spaced intervals (*FWW #65*, p.12). Bishop recommended wrapping tape around a disc to determine the circumference, then unwrapping the tape, laying out the intervals on the tape and rewrapping the tape to transfer the intervals to the disc. The question Bishop didn't answer is once you've got the tape off the disc, how do you divide it into the desired number of intervals?

First draw a series of parallel equidistant lines across a large sheet of paper. The distance between the lines should be slightly less than the smallest interval you will normally use. Number the lines starting with line 0, 1, 2, 3 and so on. Now wrap the tape around the disc, mark where the ends overlap and remove the tape. Assuming the number of desired intervals is seven, for example, you would lay the tape diagonally across the sheet so that line 0 intersects one end of the tape and line 7 intersects the mark denoting the circumference. Now mark where each intermediate line intersects your tape for accurate equidistant intervals without measuring.
—*Kathleen Wissinger, Elkton, Va.*

Double-duty marking gauge

To double the usefulness of a marking gauge, install a pencil in a screw-tightened hole at the unused end. There are many situations where a pencil line is preferable to a scratch. One can also put an India-ink drafting pen in the hole and draw nice smooth lines parallel to an edge—they look very much like ebony inlay.

—*Simon A. Watts, San Francisco, Calif.*

Scribing large circles

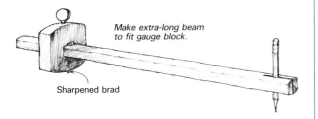

Make extra-long beam to fit gauge block.

Sharpened brad

Here's an inexpensive and easy-to-make alternative to trammel points for scribing large circles or arcs. First cut a long hardwood beam to fit the hole in your marking gauge. Make a vertical sawcut in one end of the beam, and drill a $\frac{9}{32}$-in. hole through the cut to hold a pencil. Now, drive a brad in the bottom of the gauge to serve as a compass point, install the beam in the gauge block and you're ready to scribe a circle as big as the beam.

—*Gregory V. Tolman, Evergreen, Colo.*

Trammel heads for occasional use

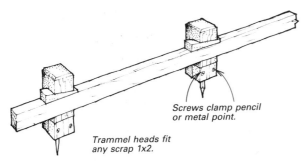

Screws clamp pencil or metal point.

Trammel heads fit any scrap 1x2.

Faced with the occasional need to draw large-radius circles, I made a pair of wooden trammel heads that will fit on any available scrap 1x2.

Each head is $1\frac{3}{4}$ in. sq. in section, with a notch to accept the beam and a wedge for tightening. A clamping block at the bottom of each head will take a steel spike or a pencil. To make the clamping block, drill a $\frac{5}{16}$-in. hole in the center of the bottom, then cut away a portion of the head halfway through the hole. Drill the block for two tightening screws.

—*Percy W. Blandford, Stratford-upon-Avon, England*

Octagon marking gauge

Many craftspeople know the traditional method for marking a square to make an octagon: First draw diagonals as shown in the sketch below. Then, with a compass set to one-half the diagonal, draw arcs from two corners. Reset the compass and walk it around the square to mark the corners of the octagon.

Repeating this procedure for different-size workpieces can be tedious. So here's a gauge, borrowed from boatbuilding sparmakers, that will scratch the lines you need along the length of a square workpiece of any width (less than its capacity), even if the workpiece tapers.

To make the gauge, first cut a cardboard square equal to the largest section you expect to deal with. On the square, draw diagonals and arcs to locate the two scribe points, as shown in the sketch. From a stout piece of hardwood make a U-shaped gauge body to fit over the cardboard square.

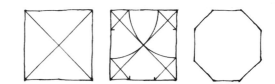

Scribe diagonals; then compass shows true octagon.

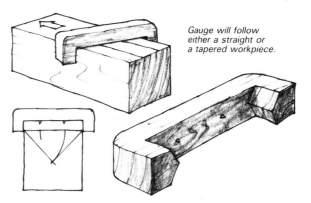

Gauge will follow either a straight or a tapered workpiece.

Drive nails in the gauge at the proper locations and sharpen. To allow the gauge to be used for smaller work, cut the ears into a prow shape as shown in the drawing.

To mark the square workpiece, angle the gauge until it bears against the sides and draw it along. If the wood tapers, the angle of the gauge will change but the proportions of the spaces across the wood will remain correct.

—*Percy W. Blandford, Stratford-upon-Avon, England*

Layout procedure for routing dadoes

Here's my pet method for routing dadoes in plywood. First locate and mark out the dado on the workpiece and score the veneer with a sharp knife. Set a compass to the distance from the edge of your router base to the bit. For example, if your router base is 6 in. across and you're using a $\frac{3}{4}$-in. bit, set the compass to $2\frac{5}{8}$ in. With the compass point on one edge of the dado, swing two arcs—one at each end of the dado. Now clamp a straightedge tangent to the arcs to serve as a fence and you're ready to rout.

—*Chuck Anderson, Porterville, Calif.*

Large-diameter caliper

Two framing squares can be used for measuring and gauging large diameters on lathe pieces. To use the squares as a fixed-size gauge, clamp them together as shown in the sketch. The framing-square gauge is actually more rigid than a large caliper.

To use the squares for measuring, leave off the clamps and simply slide both squares until they bracket the work. Then you can read the diameter on the inside scale on the back of one of the squares. With a standard 16-in. by 24-in. square you can measure up to 28 in. easily.

—*Alan Dorr, Chico, Calif.*

Adjustable protractor

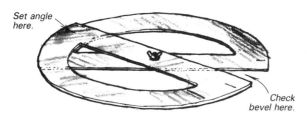

Set angle here.

Check bevel here.

An accurate adjustable angle gauge can be made quickly and inexpensively from two identical dime-store plastic protractors. With a tiny bolt and wingnut, fasten the two protractors together by enlarging the holes already made at the center. I find the device quite useful when cutting angles and also for checking the bevel angles on chisels and turning tools.

—*John Roccanova, Bronx, N.Y.*

Hole-spacing tool

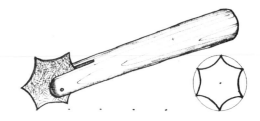

With this adaptation of a leather-stitcher's spacing tool you can quickly lay out a row of evenly spaced holes with surprising accuracy. First set a compass to the distance desired between holes and scribe a circle on 14-ga. sheet metal. Use the same compass setting to scribe six equidistant points around the circumference of the circle. Draw arcs between the points of the hexagon to create the six-pointed star shape shown in the sketch. Cut the star from the sheet metal, sharpen the points and mount the tool in a slotted handle using a nail for an axle.

—*Sandor Nagyszalanczy, Santa Cruz, Calif.*

Lathe layout tool

More convenient than a marking stick and pencil, this scribing gauge speeds spindle turning by scoring several layout lines at once. I use drywall screws as marking pins. Made from hardened steel, their tips stay sharp for making clean, thin lines. Space the screws to correspond to key measuring points on the workpiece. The gauge shown here might be used to mark divisions on a short honey dipper, but there is no reason you can't make it the full length of long work.

—*Galen Miller, Vestal, N.Y.*

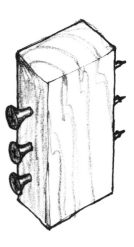

Miter gauge for plywood edgebanding

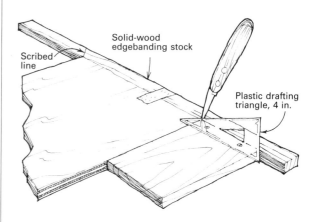

Solid-wood edgebanding stock

Scribed line

Plastic drafting triangle, 4 in.

Solid-wood edgebanding is often used around plywood doors and tabletops to cover the edge laminations. But measuring, aligning and scribing the eight 45° miters on the banding is a tedious job, because the banding's length and the miters' angles must be perfect for everything to fit. This little gauge eliminates the measuring and allows you to mark the miters right from the workpiece.

To use the jig, tape the banding stock in place on the edge. Slide the jig into the corner where the work and the banding stock meet, and scribe the 45° miter with a sharp knife. Move to the other end and repeat. If the jig has been accurately made, you'll have perfect scribe lines for cutting miters on the banding.

To cut the miters, I use a standard plywood jig with rails on the bottom that run in the miter-gauge slots on my tablesaw.
—*L. A. D. Colvin, Satellite Beach, Fla.*

Sliding measuring sticks

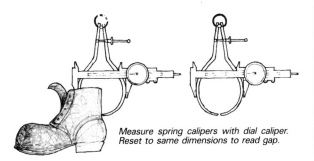

Adjust to inside dimensions and clamp together or mark with pencil.

Tongue-and-groove glue joint

The next time you're running tongue-and-groove glue joints, rip off some sticks about ⅜ in. thick, with the joint profile on one side. Put the two sticks together to make a sliding measuring stick that can take inside measurements accurately. The measurements can be registered by marking across both sticks with a pencil or by clamping the two sticks together with a small C-clamp. The measuring stick is especially useful for checking a door opening for consistent width from top to bottom.
—*Robert M. Vaughan, Roanoke, Va.*

Measuring wall thickness in carvings

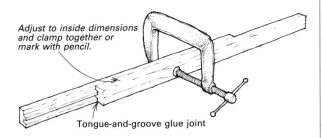

Measure spring calipers with dial caliper. Reset to same dimensions to read gap.

While carving a wooden shoe, I wanted to measure the wall thickness near the ankle. Since none of the measuring tools I had would do the job directly, I used the two-step technique shown in the sketch above.

If you were measuring the thickness of several spots, it would pay to make up a table of wall thicknesses and blade-spread measurements beforehand, so you wouldn't have to reset the calipers each time. I suspect this method would be useful for measuring wall thicknesses on hollow turnings as well.
—*Gilbert J. Warmbrodt, St. Louis, Mo.*

Two wall-thickness calipers

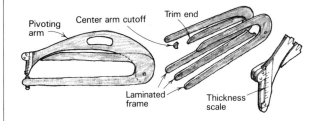

I believe this thickness caliper is simpler and less prone to error than Gilbert Warmbrodt's technique, which involves using both a dial caliper and a spring caliper. I made mine from plans carried in a 1950s British woodworking magazine.

The calipers consist of a 3-piece laminated frame and a pivoting arm. Make it by screwing together three pieces of ³⁄₁₆-in. Baltic birch plywood (sold in model and hobby shops). Next, cut the three pieces into the U-shaped caliper frame. Cut the pivoting arm from Baltic birch to the shape shown in the sketch. Disassemble the three frame pieces and cut apart the middle piece as shown to allow clearance for the pivoting arm at the back and front of the caliper. Save the cutoff from the center piece; the pointer plate is screwed to it and then attached to the left frame piece. Trim the front of the right frame piece so you can read the measured thickness.

Glue the three frame pieces together, bore a hole through them and bolt the pivoting arm in place with a wing nut on its end to adjust the pressure on the arm. For the scale, mark off a strip of ¹⁄₁₆-in.-thick aluminum or brass in ¹⁄₃₂-in. increments. Screw or glue the plate to the pivoting arm as shown. Sharpen the end of a small bolt and center it on the bottom of the frame so the tip of the pivoting arm meets it. With the caliper closed, mark a fine line from the thickness scale to the pointer plate to indicate zero.
—*John Bickel, Ossining, N.Y.*

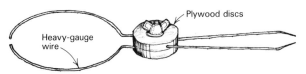

These shopmade calipers are made from two 3-in.-dia. plywood discs, a bolt and wingnut and four pieces of heavy, stiff wire. Accuracy depends on two conditions. First, the distance from the pivot to both ends must be exactly the same. Second, when the caliper jaws are closed, the two chisel faces at the other end must also touch. To use, bring the curved ends together on the workpiece and measure the thickness as the distance between the two chisel faces on the other end.
—*Ralph S. Mason, Portland, Ore.*

Ellipse layout revisited

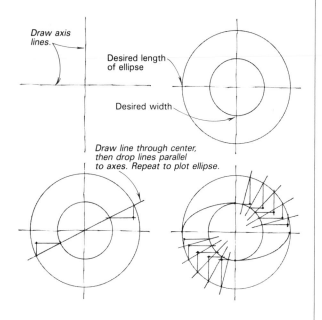

Draw axis lines.

Desired length of ellipse

Desired width

Draw line through center, then drop lines parallel to axes. Repeat to plot ellipse.

When laying out an ellipse, most people care more about its finished length and width than about the distance between the two focal points. The draftsman's method shown in the sketch gets directly to the point without requiring calculations and gadgets.

—*Lawrence Whytock, Brockville, Ont.*

Drawing a half-ellipse

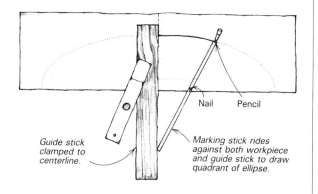

Nail   Pencil

Guide stick clamped to centerline.

Marking stick rides against both workpiece and guide stick to draw quadrant of ellipse.

We use this method to lay out seat backs in our shop. First mark the rise and run of the half-ellipse on the stock and clamp a 1x2 guide stick on the centerline, as shown in the sketch. Now cut a 1x1 marking stick half as long as the ellipse and notch one end to hold a pencil point. Drive a nail through the stick at a distance from the notched end equal to the rise of the ellipse. To draw the ellipse, hold a pencil in the notch and move the head of the stick from left to right while riding the nail against the stock and the tail of the stick against the guide stick. Reclamp the guide stick to the other side of the centerline to complete the curve.

—*Doug Hansen, Letcher, S. Dak.*

Drawing giant shallow arcs

Drawing a large-radius arc through three points isn't easy if you don't have room to use a rope as a compass. Here's a method I worked out while arranging some permanent chairs.

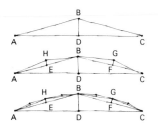

If A, B and C are points on the circle, first connect these points and drop a perpendicular from B. At the midpoints of AB and BC draw two perpendicular lines and measure along them a quarter of the distance BD. Repeat the operation with the new outside lines, this time measuring a quarter of the distance FG. Continue the process until you have a close approximation of the true arc. It will be surprisingly accurate—especially if the arc is only a small portion of the circumference (i.e., BD is small compared to AC).

—*Christopher Yonge, Lothian, U.K.*

Drawing large shallow curves

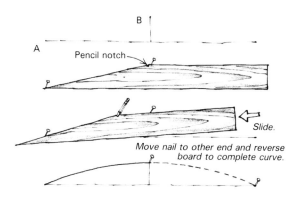

B

A

Pencil notch

Slide.

Move nail to other end and reverse board to complete curve.

When I was a boatbuilder we used this shallow-curve drawing method to set out the deck beams of yachts. The trick works for drawing any such curve with a known rise and run.

You'll need two nails and a "spile board." Cut the spile board as wide as the curve's rise and taper the board on one end with the length of the taper equal to the curve's run. Notch the board at the location shown to catch a pencil point.

Drive one nail at point A and another at point B. With a pencil in the notch and the spile board positioned as shown in the sketch, slide the board toward the nail at A to draw the curve. Nail A can be removed and driven in the other end to complete the curve. In our situation the method was used to make a template from which all the shorter beams and carlings could be marked.

—*Ernie Ives, Ipswich, England*

Dowel-center marker

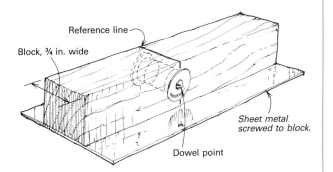

Reference line

Block, ¾ in. wide

Sheet metal screwed to block.

Dowel point

If you use a lot of doweled edge joints, this dowel-center marker will speed up the layout process by marking the mating holes simultaneously. It's made from two steel dowel points, a hardwood block and a small piece of sheet metal. First, countersink for the shoulder of the steel points on opposite sides of the block, and then drill the through hole. If you have access to a drill press, use it to ensure accuracy. Glue the points in place with epoxy, and attach the sheet-metal fence with a couple of small flat-head screws.

Lay the two boards, aligned just as they're to be glued, on a clean workbench. With a framing square, draw a line across the two boards where a dowel will be needed. Now, place the dowel-center marker between the boards, matching the reference line on the marker block with the lines drawn on the boards. Tap the boards together, against the points, to mark both boards at the same time.

With long material, I mark and drill the first mating holes on one end and join the pieces dry with a dowel ¾ in. longer than I intend to use. This keeps the boards in registration and holds them far enough apart so I can move the marker to any point along the length of the boards and make accurate marks.

—*Harry E. Hunter, Oakville, Conn.*

Center finder from a corn-chip can

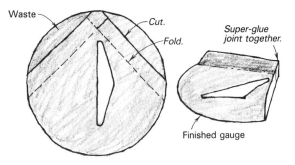

Waste

Cut.

Fold.

Super-glue joint together.

Finished gauge

This simple, handy tool for spindle turners pinpoints the headstock and tailstock centers on round, square or octagonal blanks. To make the gauge, first tear off the circular aluminum top from a 15-oz. can of corn chips or other snack food. The thin aluminum disc is the right size and can be cut with scissors or, if backed up with a hardwood block, by a sharp wood chisel. Scribe, cut and bend the disc as shown in the sketch to produce an L-shaped lip and a diagonal marking opening. To ensure an accurate center, mark at least four diagonals on the end of the spindle and pick the point where the most lines cross.

—*Eliot Birnbaum, Syracuse, N.Y.*

Homemade center finder

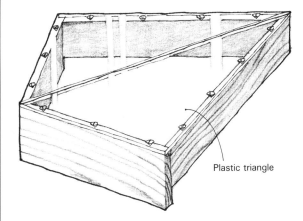

Plastic triangle

To make this inexpensive but accurate center finder, purchase two identical 30°-60°-90° triangles at an art or drafting-supply store. The triangles are available in a variety of sizes. Attach the two triangles to a wooden frame using small brass screws. Leave a ¹⁄₁₆-in. space between the two triangles to provide a marking slot. On my center finder I made the frame deeper on the 60° end, as shown in the sketch, so I could use the center finder for marking objects too large to fit within the frame.

—*Dave Sander, Port Orchard, Wash.*

Gluing and Clamping

Chapter 6

Gluing coopered panels

Here's a fast way to glue up staves for a coopered panel. Apply the glue to as many as a quarter-circle's worth of staves (if the panel is bigger than a quarter-circle, glue it up in sections). Then, lay the staves edge to edge, outside-up, on a clean flat surface. Apply several strips of strapping tape (the kind with fiberglass filaments running the length) across the staves, taking care to keep the stave edges in close contact. Now, using two pipe clamps on the inside of the curve, apply light pressure to close the gaps and hold the panel in its curved shape until the glue sets. The procedure sounds too easy, but I've made strong panels with invisible glue lines inside and out using the method.

—*Gregory V. Tolman, Evergreen, Colo.*

Tinting glue

Since most carpenter's glues dry almost colorless, any squeeze-out is hard to see until you apply the stain and see that telltale white spot. To solve this problem, tint your glue with ordinary food coloring. A few drops of red or green will make any squeeze-out highly visible and easy to sand off. If your taste in color is more conservative, mix equal amounts of red and green to make a pleasant brown.

—*Donald F. Kinnaman, Phoenix, Ariz.*

Improved glue spreader

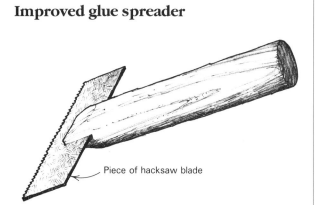
Piece of hacksaw blade

I've seen several ideas for glue spreaders in "Methods of Work," all of which, in my opinion, have a serious drawback: Their smooth edges make it difficult to spread a consistent amount of glue over a broad surface, because the slightest variation in pressure varies the amount of glue on the surface.

To avoid this problem, I use broken or worn-out hacksaw blades for glue spreaders. The toothed edge works just like a tile layer's notched trowel, leaving a consistent layer of glue wherever it's spread. To make a handle for the spreader, I hacksaw a short slot in the end of a dowel scrap and press the section of blade into the slot.

—*David W. Engel, Joliet, Mont.*

Mason-jar glue pot

A mason jar makes a good glue pot. Drill a hole through the lid insert to fit the handle of a disposable foam paint-brush. To let the brush hang in the pot, put the insert ring on top of the twist-on cap, as shown in the sketch. You can adjust the height of the brush with a rubber band around the brush handle. The twist-on ring gives a nice scraping edge for wiping excess glue off the brush. —*David L. Pitz, Redding, Calif.*

Eliminating glue squeeze-out

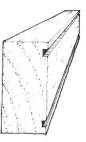

I discovered this interesting solution to an ancient gluing dilemma while restoring an old drawer. The drawer's guide was glued in place, but there was no glue squeeze-out to be seen. The maker had sawn two shallow sawkerfs into the gluing surface near the edges. When he applied glue to the center section of the guide and clamped it in place, any potential squeeze-out was contained in the kerfs.

—*John M. Gray, Syracuse, N.Y.*

Glue injector

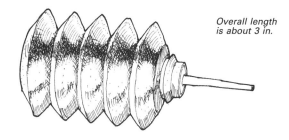
Overall length is about 3 in.

Ask your veterinarian to save you a few of these little accordion squeeze bottles that come filled with an antiseptic used to irrigate puncture wounds. The bottles make great glue or oil applicators in tight places. —*Steve Allard, Carbondale, Ill.*

Hot-melt adhesive by the sheet

When I couldn't find sheets of hot-melt adhesive for my veneering project at the store where I bought the veneer, I went looking elsewhere and found it at a fabric shop. The sheet adhesive comes in two forms: fusible web and transfer web. Fusible web is simply a sheet of hot-melt adhesive that you cut to size, place between the parts and press with a hot iron to fuse. With this type, there may be a bit of squeeze-out around the edges of the veneer.

By contrast, the transfer-web adhesive is applied by first melting the adhesive to the back of the veneer. Then you peel off the backing paper, cut the veneer to size and fuse the veneer to the ground with a hot iron. This approach seems to result in a neater glue job. To avoid ironing dirt into the wood, place a clean sheet of paper between the iron and the veneer. A sandbag will apply the small amount of pressure needed to hold the veneer in place while the glue sets.

—*Gerald W. Edgar, Renton, Wash.*

Iron-down veneer

To apply veneer edging strips, coat the strip and the edge with white (not yellow) glue. My glue applicator is a plastic-laminate roller covered with masking tape, which I remove after the glue is spread. Allow the glue to dry. Then iron down the veneer using a household iron on the highest setting. Move the iron back and forth over a 12-in. to 14-in. area to heat it, then press with the roller to fix the bond. This method produces a strong glue joint with no mess, no clamps, and no fumes from contact cement.   —*William J. Bosco, Garberville, Calif.*

Yet another clamp

These clamps, which are patterned after a set of steel ones made

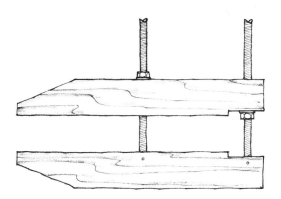

years ago by the grandfather of a fellow worker, are inexpensive, easy to build and suit my small work better than purchased models.

Construction is straightforward. Select a strong, hardwood such as oak or maple and cut two identical jaws. Clamp the two jaws together and drill the two threaded rod pilot holes. The holes should run through the top jaw and about an inch deep into the bottom. Enlarge the through holes in the top jaw to slip over the threaded rod. Although it is not necessary, you may wish to cut steps at the backs of the jaws, as shown, so that the jaws can be tightened down to zero. Tap the holes in the bottom jaw and screw in the lengths of standard threaded rod. Next drill a 1/16-in. pin hole through the cheeks of the jaw and through the threaded rod. Pin the threaded rod in place with a finish nail. Ordinary washers and nuts installed as shown in the sketch finish the clamps. For large clamps you can buy "Quick Acting Hand Knobs" from Reid Tool Supply Co., 2265 Black Creek Road, Muskegon, Mich. 49444 (part No. QK1, $3.26 each). When these knobs are tilted, they slide freely along 1/4-20 threaded rod, allowing quick adjustment.

To use the clamps, first adjust them to the work, keeping the jaws nearly parallel. Tighten the outside nut gently and use the inside nut to apply pressure. Use a light touch. These clamps can develop awesome force with just a few turns of the screws.

—*Raymond Levy, Soquel, Calif.*

T-nut handscrews

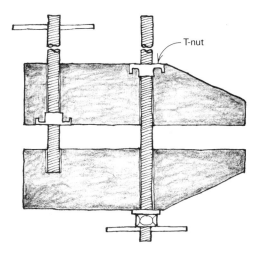

For a few cents worth of hardware you can transform a couple of pieces of scrap into durable small handscrews. The sketch shows the general construction idea. Handles could be turned from wood, but I just use T-bar handles made from 1 1/2-in.-long pieces of brazing rod that press-fit into 3/32-in. holes in the threaded rod. Note the nut and washer on the front screw (to leave space between the handle and the clamp) and the flat bottom on the back hole (to prevent the threaded rod from splitting the clamp). I like to counterbore the T-nuts for a neater appearance and put a dab of epoxy under them to make sure they stay in place.   —*Chuck Anderson, Porterville, Calif.*

Picture-framing clamp

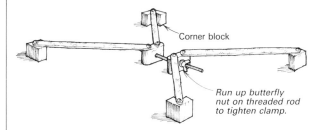

For clamping picture and mirror frames, I have made several sizes of the jig shown above. In use, the corner blocks fit over the mitered ends of the molding. When the wing nut is tightened, equal pull is placed on the four arms, which pull the miters equally together. For best results, when the clamp is tight, its arms should be at about 90° to each other. This angle depends on the length of the arms in relation to the size of the frame; hence, the various sizes. For frames of unusual proportions, a few assorted lengths of 3/8-in. threaded rod give me all the range of settings I need.   —*C. Robson, Coe Hill, Ont.*

Miter clamping cleats

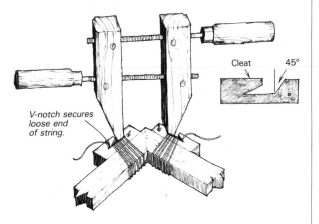

V-notch secures loose end of string.

Cleat   45°

Here's an easy way to attach clamping cleats for gluing mitered joints. You'll need eight of the clamping cleats, which can be cut out on the bandsaw in just a few minutes. The cleats can be made up in any size using about the same proportions as those in the sketch. They should be about the same thickness as the stock being glued.

To use the cleats, set one in position near the corner to be glued and wrap the attached string around the frame six or eight times. Secure the loose end of the string in the V-notch of the cleat. Repeat with other glue cleats until you have a pair installed at each corner. Now, pull the cleats together with clamps. The cleats will move a fraction of an inch at first and the string will creak as it takes the strain. But, soon, the cleats will hold tight, giving your clamp a perfect perch to draw the joint tight.

—*David Wardale, Merced, Calif.*

Power wedges for edge-gluing

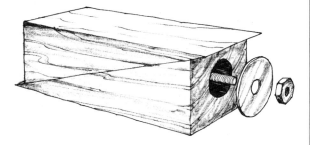

Small wedge pairs are a good way to clamp thin stock when edge-gluing (see *FWW* #58, p. 45) because the wedges won't exert too much pressure. But recently I needed to clamp a glue joint where wedges were the only thing that would fit, yet I had to apply a lot of clamping pressure to the joint. To solve the problem I came up with the power wedges shown above. These wedge pairs work just like the pairs you tap into place. But since they are tightened with a machine screw, you can apply much more clamping pressure.

Start with a rectangular block about 1 in. by 1 in. by 3 in. Drill an oversize hole lengthwise through the block, cut the block in two on the diagonal, then add a machine screw and washers as shown.

—*Richard Farwell, San Luis Obispo, Calif.*

Clamping hexagonal box tops

I enjoy making small hexagonal boxes because they are a greater challenge to construct and are more visually interesting than square boxes. Gluing up six pieces for the tops, however, presents a problem. The jig pictured here solves things by securely holding all the pieces with one bar clamp.

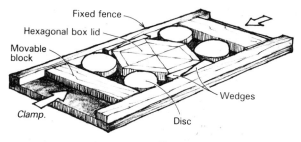

Fixed fence

Hexagonal box lid

Movable block

Clamp.

Wedges

Disc

The jig consists of a plywood tray with two parallel fences fastened to the long edges. The operating width of the tray is adjusted by means of two pairs of wedges, as shown. The bar clamp spans the two clamping blocks, and four wooden discs redirect the pressure at the proper angles.

—*Steven Barnhill, Gunnison, Colo.*

Clamping round tabletops revisited

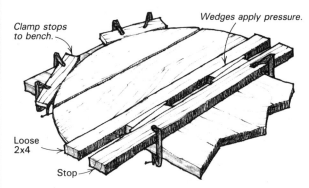

Clamp stops to bench.

Wedges apply pressure.

Loose 2x4

Stop

Here's an alternative to the clamp-perch idea (*FWW* #47) for gluing circular tabletops. First place the tabletop on the bench and clamp three stops as shown in the sketch above. Place a free-floating 2x4 against the workpiece and drive paired wedges between the clamped and floating 2x4s to apply pressure to the glue joint. Here are some additional tips: Place newspaper on the bench to catch the glue that will squeeze out, and dowel the edge joint to keep the pieces from shifting. It's best to raise the tabletop up with a few thin scraps of wood—this centers the clamping pressure and allows glue to drip out without smearing. I put weights on top of the work if necessary to keep it flat while the glue sets. And as with any glue-up, it's a good idea to make a dry run first.

—*Ken Jones, Lisle, Ill.*

Clamping perch for irregular shapes

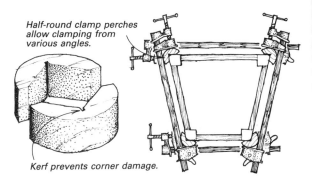

Half-round clamp perches allow clamping from various angles.

Kerf prevents corner damage.

To clamp up chair frames and other irregular shapes, I use semicircular clamping perches like the ones shown in the sketch. To make the perch, cut a 3-in.-dia. circle from a 2x4 with a large hole saw or fly-cutter. On the bandsaw, halve the circle with the grain and make a V-cut in the flat side to match the angle of the corner of the piece being glued. An extra kerf at the apex will prevent the block from crushing any sharp edges or from being glued to a mitered corner. The round surface will accept clamping from any angle, even cross-clamping.

—*John M. Gray, Syracuse, N.Y.*

Universal bending form

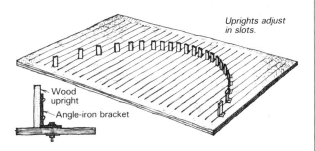

Uprights adjust in slots.

Wood upright

Angle-iron bracket

I was recently commissioned to build an arch for a client who had knocked an opening in the wall between his kitchen and dining room. Not wanting to be stuck with a big, expensive form when the job was done, I made an adjustable and reusable form for bending the arch.

The form was made from 3/4-in. plywood, and was through-slotted every 4 in. with a 1/4-in. bit in my plunge router. I stopped the slots about 2 in. from each edge. I then strengthened the plywood (made flimsy from the slots) with a 1x2 wooden frame and three crossmembers attached to the bottom. I had a local metal shop make 25 angle-iron brackets, to which I attached pine blocks whose edges had been rounded over to minimize marking the work. A 1/4-in. bolt with a small washer on top and a large washer underneath holds the bracket to the formboard after it has been set at the proper location. If a bracket is needed where there's no slot (as often happens near the ends of the curves), I just drill a 1/4-in. hole where the bracket's needed.

—*Jason Tessler, D.N. Maalay Hagaziz, Israel*

Extending pipe clamps

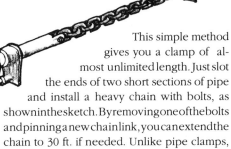

This simple method gives you a clamp of almost unlimited length. Just slot the ends of two short sections of pipe and install a heavy chain with bolts, as shown in the sketch. By removing one of the bolts and pinning a new chain link, you can extend the chain to 30 ft. if needed. Unlike pipe clamps, which must be flat to work, the chain will bridge minor obstacles without loss of pull. An added bonus is that the chain requires little storage space.

—*Harold R. Olsen, Fox Island, Wash.*

Reversing pipe clamps

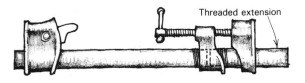

Threaded extension

It's handy to be able to reverse a pipe clamp so it can be used to push something apart. In fact, special clamp heads are sold for this purpose. As a thrifty alternative, if you add a short section of pipe to the head as shown, you'll be able to reverse any standard pipe clamp at will.

Screw the head on backwards and stop about halfway. Now screw the short 6-in. piece into the clamp head in the normal fashion. Reverse the shoe, and you have an efficient spreading clamp.

—*T.D. Culver, Cleveland Heights, Ohio*

Preventing clamp stains during glue-up

Bandsaw wax-paper roll, then tear off strip to cover clamp bar.

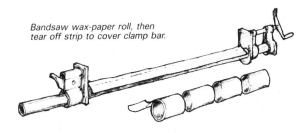

Here's a simple solution to stains caused by wet glue reacting with metal clamps. Bandsaw a roll of ordinary wax paper into 2-in. mini-rolls. Tear off what you need, fold into a tent shape and lay the wax paper on the clamp as shown.

—*Dustin P. Davis, Frostburg, Md.*

Insulating pipe clamps

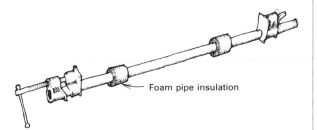

Foam pipe insulation

To prevent glue stains and dents while using your pipe clamps, cut two or more 2-in. sections from a length of foam pipe insulation and install the sections on the pipe as shown in the sketch. Foam pipe insulation is commonly available in several sizes at plumbing and building centers.

—*Alan C. Sandler, Garnerville, N.Y.*

Homemade edge-gluing clamps

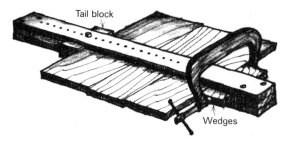

Tail block

Wedges

Here's an inexpensive but effective homemade clamp for edge-gluing stock. Unlike a pipe clamp, it won't fall off the workpiece while you're fitting up and it pulls evenly on both sides of the stock, ensuring flat panels. To use, pin the sliding tail block in an appropriate place, then apply pressure by screwing down a C-clamp across the wedges. Scraps of wax paper will shield the clamp from glue squeeze-out.

—*Bert Whitchurch, Rockaway Beach, Mo.*

Preventing panel clamp-up buckle

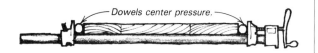

Dowels center pressure.

When clamping up boards to make a panel, pipe clamps can ride up on the fixed jaw, causing the pressure to be uneven and the panel to buckle. This is caused by the angle of the jaw changing as the pressure comes on. If this bothers you, take a dowel having the same diameter as the thickness of the panel and lay it between the jaws and the work. Now the pressure will be applied on the center of the panel edge just where it should be. The dowel will likely dent the edge, so you may want to add a piece of scrap as a buffer.

—*Henry T. Kramer, Somerville, N.J.*

Double-duty edge-gluing clamps

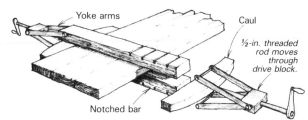

Yoke arms

Caul

½-in. threaded rod moves through drive block.

Notched bar

This shop-built edge-gluing clamp performs double duty. It not only is a terrific bar clamp, but it also aligns the various workpieces being glued, thus eliminating the need for a separate alignment "sandwich" made with scrap and C-clamps.

The clamps consist of two yokes and two notched wooden bars. Each yoke assembly has a pair of trapeze-like arms made from 8-in.-long pieces of strap iron that pivot on the sides of a block made from ¾-in.-thick mild steel. Drill and tap a hole through the block to accept a ½-in. threaded drive rod. Then, drill and tap ¼-in. holes in the sides of the block to bolt the arms in place.

Next, screw a length of ½-in. threaded rod through the block and attach a knob or crank to its outboard end. To distribute clamping pressure, make a wooden caul with a shallow ½-in. hole bored in its edge to locate the end of the rod. Plane a shallow concave curve in the caul edge that contacts the work to ensure even distribution of clamping pressure.

Cut the clamp's notched wooden bars from 1¼-in.-thick sticks of hardwood. The bars should be as wide as the space in the yoke arms. To make sure the notches in the bars are perfectly aligned, cut both bars at the same time with a ¼-in. dado blade

—*William Swartz, Modesto, Calif.*

Edging plywood with pneumatic clamps

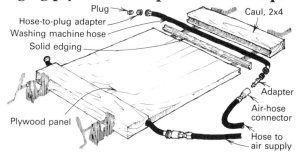

Plug

Hose-to-plug adapter

Washing machine hose

Solid edging

Caul, 2x4

Plywood panel

Adapter

Air-hose connector

Hose to air supply

Clamping solid wood edging to plywood panels with standard bar clamps is not only a tedious and time-consuming job, but it also depletes my entire supply of clamps. So I designed a pneumatic clamping system that is faster, gives strong, even pressure to the wood and results in virtually invisible gluelines.

The system's key component is a 2x4 caul, coved along one edge to fit around a stiff, high-pressure hose. Washing-machine connector hose is rated at 125 psi and works well. Plug one end of the hose and fit the other with a standard air-hose connection.

To use the device, clamp it to the plywood being edged with a couple of bar clamps. Center the hose over the edging being glued. Then connect the air supply (60 to 100 psi) to activate the pneumatic clamp and watch the gaps disappear as the hose expands. I made my cauls 10 ft. long so I can clamp three cabinet sides at a time. To conserve space, I plan to mount the units on the wall and stack the panels vertically. —*Jeffrey P. Gyving, Point Arena, Calif.*

Joinery

Chapter 7

Hybrid paneling system

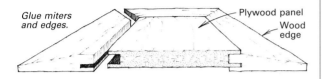

Glue miters and edges.

Plywood panel

Wood edge

On a recent project that called for Georgian wainscotting on and around a stairway, I devised a way of combining solid-wood edges with ½-in.-thick plywood centers to produce large fielded panels. The approach takes advantage of the superior stability and affordability of plywood while avoiding the unsightly glue layers and voids that show on the beveled edges of all-plywood panels.

To make the panels, first cut all the plywood centers to size. Then make the solid-wood bevel stock for the panel edging. This can be done easily with a thickness planer and a shopmade bed to tilt the stock sideways a few degrees.

The panel centers are fitted to the bevel frame with a double tongue-and-groove joint. To produce this joint, use a ¼-in. slotting cutter on your router table to groove the edges of the plywood panels and a ⅛-in. cutter to groove the edge of the solid-wood bevel stock. The trick in routing the slots is to set the height of the cutter so you leave a ¼-in. tongue on the bevel stock, which will press easily into the slot in the plywood. And conversely, the tongue left on the plywood should press into the groove in the bevel stock.

Now carefully miter the bevel pieces, apply glue to the tongue and tap the frame gently into place around the panel. You may wish to pin each corner with a brad, but there's no need to clamp the assembly if the tongue-and-groove joints fit correctly.

I used this technique not only for the rectangular wainscotting panels, but also for the parallelogram-shape panels and triangular panels at the side of the staircase. My work was to be painted, but I see no reason why this technique would not look fine with a stained finish.

—William D. Lego, Springfield, Va.

Perfect half-lap joints

The half-lap joint is a strong, useful joint for cabinet frames, but setting up the saw so it cuts away exactly half from each piece can be frustrating. Here's a method for setting the blade height

that's fast and foolproof, and it doesn't require jigs or measuring. Grab a waste piece of the frame stock, and with your tablesaw's blade height elevated to less than half the stock thickness, pass the stock over the saw on both sides to leave a thin center section. Raise the blade a little and make a second pass. Continue raising the blade, taking cuts on both sides, until only the slightest membrane of wood remains in the center. You have now achieved the perfect blade-elevation setting. I waste the lap area with repeated passes, then clean up any small ridges and saw marks with a rasp.

—Duane Waskow, Cedar Rapids, Iowa

Making finger joints

The finger joint is not only an effective corner joint, but it can also be used for sharp bends and curves. This method for making finger joints minimizes cumulative error. I stack up four identical 6½-in. blades on my tablesaw with spacers between them. The spacers must be made to a prescribed thickness so the slots are the same width as the fingers. To determine spacer thickness, first measure the tooth width and the blade thickness with a micrometer. To calculate the spacer thickness, double the tooth width and subtract the blade thickness. The spacers can be made from items normally found around the shop, such as Formica. Wafers cut from thin sheet metal or soda cans make good fine-adjustment shims.

—Kenneth T. May, Jeanerette, La.

Veneering end edges

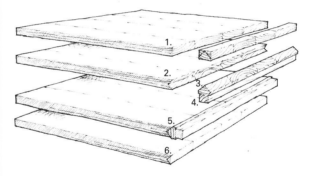

I use this process when building veneered period reproduction pieces that have a solid lumber core. This technique provides a better gluing surface than endgrain for the veneer without creating expansion problems as a breadboard end would. Step 1: After gluing up the core, slice a strip off each end that is ¹⁄₁₆ in. wider than the core's thickness. Step 2: Cut a V-notch in the end of the panel. I do this on the shaper, but you can also do it on the tablesaw. Step 3: The strip will have two end-grain edges and two face-grain edges. Pick a face-grain edge and saw a peak on it as shown in the drawing. Step 4: Glue a scrap strip to the piece to act as a caul. Step 5: Glue the pointed end into the notch. Step 6: Saw off the caul, and using a handplane, fair down the glued-on piece so it is even with the core's edges.

—Harold Ionson, Westwood, Mass.

Simplified captured-nut system

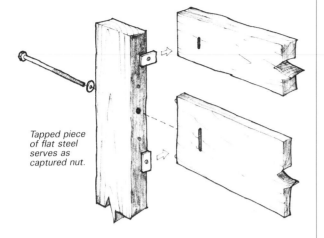

Tapped piece of flat steel serves as captured nut.

I'm familiar with the principle of barrel nuts—short lengths of round steel rod, drilled through and tapped—used as captured nuts to join stretchers to frames. Unfortunately, when I tried to manufacture them, I quickly discovered that the metalworking tools and skills required were beyond my meager means.

Instead, I returned to my own system, shown in the sketch. It uses a piece of flat, 1½-in.-long, ¾-in.-wide steel that is drilled, tapped and fitted into a routed slot in the stretcher.

The rectangular shape of the "nut" allows maximum purchase, and you can vary its orientation as shown, depending on the thickness of the frame members.

—*Chuck Lakin, Waterville, Maine*

Plug locates nut

Embedded nuts are a convenient joinery technique for beds, trestle tables and other projects that may have to be knocked down for moving or storage. The method shown here, in which a square nut is held in place by a slotted dowel plug, has some nice features and can be adapted to a variety of sizes. My usual combination is a standard bedbolt and nut (available from most period hardware suppliers) with a 1-in.-dia. dowel. For clarity, the drawing omits the stub tenon on the rail and the mortise in the post, which, if you use just one bolt, are necessary to keep the joint from twisting.

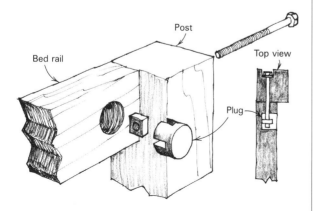

Post

Bed rail

Top view

Plug

I slot the end of the dowel by running it over a ⅜-in. dado blade in the tablesaw. I've devised holding and indexing jigs that attach to my miter gauge for this job because I make about a year's worth of plugs at one time. For just one or two projects,

you could use a regular blade and make two or three side-by-side passes until the slots are wide enough to accommodate the bolt and nut. It would be a good idea to use the end of a 2x4 scrap as a push block to keep the dowel vertical and prevent it from kicking back. The depth of the slot should allow the end of the inserted nut to be just flush with the end of the dowel. Then cut the plug off a little longer than necessary—as it will be smoothed flush after it is glued in place in the rail—and continue cutting plugs until the dowel gets too short to handle safely.

The plug gives a nice inside finish to the rail, stops the nut from turning, and prevents the nut's corners from cutting into the rail, which would eventually loosen the joint.

Alignment of the holes is critical. One trick is to drill the bolt-holes in the bed's posts on the drill press, to ensure that they are centered and straight, then, with the joint assembled, use an electric drill (with the holes in the bedposts as pilots) to continue the bolt-hole into the rail. Size the depth of the dowel-hole in the rail so that the bolt engages the nut smoothly. Test the alignment before gluing the dowel and nut in place; if you are a little off, you can enlarge the bolt-hole somewhat without weakening the joint much.

—*Christian Becksvoort, New Gloucester, Maine*

Lag screws in endgrain

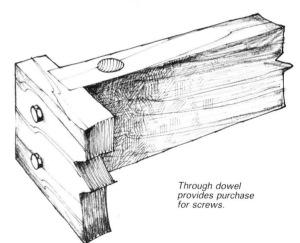

Through dowel provides purchase for screws.

Lag screws driven into endgrain are weak and won't hold under pressure. But if you drill a hole through the member and add a dowel as shown, the screw can bite into the long grain of the dowel and turn a weak dado into a strong, practical joint. I used this construction to connect the front and side rails of a knock-down pine sofa frame. —Jack Fisher, New Hope, Pa.

Aligning pins with holes in table leaves

Here is a simple method for aligning the pins in table leaves with holes in the expanding table's top. Before gluing up the tabletop and leaves, take a piece of the tabletop stock and drill the pin holes squarely through it using a drill press. Then rip this piece into several 1-in.-wide strips, enough to laminate a strip to all interior edges. All that remains is to glue up the top and the leaves with a strip on each edge where pins and holes must mate. —*Kathleen Sillick, Gasport, N.Y.*

Hidden floating-dowel joint

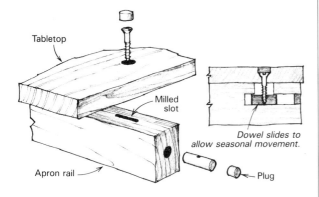

This hidden joint discretely accommodates the seasonal movement of a solid-wood tabletop in relation to its apron. It could also be used to fasten a seat to its rails or a shelf to its brackets. Start by drilling a hole lengthways into the end of the apron. Then mill a narrow slot all the way down into the hole cavity, as shown. Insert a short dowel that's loose enough to slide easily into the hole, and center it under the slot. Plug the hole in the apron and finish as desired.

To fasten the tabletop to the apron, drive a countersunk and plugged screw through the top, through the milled slot and into the dowel. If you don't want screw plugs to show on the top, the construction of the joint can be reversed with the dowel installed in the top and the screw driven through the apron from below.

—*Sandor Nagyszalanczy, Santa Cruz, Calif.*

Frame joint for a job-site table

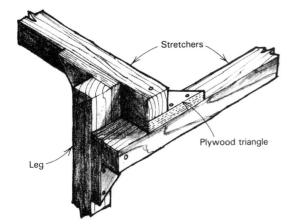

It's often useful to be able to construct a solid table or bench frame quickly at the job site. I simply cut legs and stretchers to length from scrap, then join with three 45° triangles of ⅜-in. plywood at each corner, glued and nailed. Variations of this joint—with a little mathematical figuring and diamond-section ribs—can be used for quick geodesic domes and other timber-frame structures. —*Chris Yonge, Edinburgh, Scotland*

Sliding frame-to-carcase joint

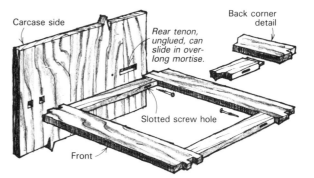

I don't claim to have invented this joint. It has surely been used before, but I have never seen it explicitly described. The joint is designed to install parting frames, such as drawer supports, into a solid wood carcase. It provides a sliding action to allow the solid-wood carcase sides to expand and contract with changes in relative humidity.

The double tenon at the front of the frame is glued into a double mortise in the carcase, locking the frame in position (a single tenon would be glued mostly to endgrain in the carcase sides, which doesn't make for a very good glue joint; the double tenon has twice as much side-grain to side-grain gluing area). At the back, a sliding tenon in an extra-long slot and a screw through a slotted hole allow the carcase sides to shrink and expand without damage.

—*Nicholas Cavagnaro, Orofino, Idaho*

Strengthening curved frame members

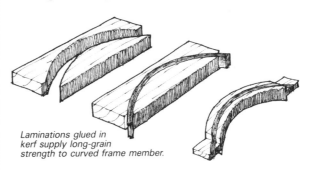

Laminations glued in kerf supply long-grain strength to curved frame member.

Curved frame members can look attractive. But when curves are cut from solid wood they may be structurally unsound, because the long grain is severed. Fully laminating the piece adds the necessary strength but requires special forms and much fussing. In addition, the laminated workpieces are difficult to machine further. Here's a procedure that solves these problems. It adds strength where needed, no forms are required, and the resulting workpiece is easy to machine.

To make the frame member, first saw the workpiece blank into two pieces, following the midline of the curve. Using three or four 1/10-in.-thick plies, glue up a sandwich as shown. When the glue sets, trim off the excess laminations, cut the member to shape and machine.

—*Jim Fawcett, Rosendale, N.Y.*

Swivel joint for coopered door

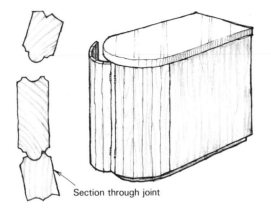

Section through joint

I recently built solid-wood cupboards for our kitchen, and planned to have one counter stand out from the wall like a peninsula. In order to avoid sharp (and potentially painful) corners, I rashly decided that it should have a semicircular end with curved doors. The problem then became how to make two coopered doors of the correct radius. I didn't want to try tapered-brick construction, because any slight error in the taper angle multiplies—in my doors, even ½° of error would have thrown the radius off by about an inch.

I eventually hit on the idea of the swivel joint shown in the sketch. This joint, which I made with a router, allows the pieces to be shifted and glued up at the exact radius needed. Note that the male part of the joint is a full half circle, while the female part is of a matching radius but shallower.

I made a temporary gluing jig from chipboard to hold the boards at the proper radius while the glue was drying. I incorporated end stops as well, at the correct width of the door, so that as the pieces were clamped to the radius they would also be pressed tightly together at the joints. During dry-assembly, I fitted the pieces into the jig and carefully planed the outer edges of the door to a perfect fit, then glued and clamped the assembly with two web clamps wrapped around both jig and door. —G. L. Degg, Stoke-on-Trent, England

Stop molding for crowned frames

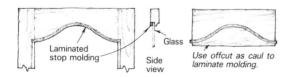

Laminated stop molding. Glass. Side view. Use offcut as caul to laminate molding.

The usual procedure for making stop molding for glazed crowned frame doors is to bandsaw the molding from solid stock. This approach presents two problems. First, it's difficult to fit the curved molding to the frame. Second, the inevitable endgrain of the bandsawn piece is weak and prone to splitting when nailed in place. Both these problems can be eliminated by using laminated veneer strips to make the molding. I cut the strips slightly oversize in width, and use the actual door and its waste piece as a two-part form to shape the wetted and glue-coated veneer strips. After the glue has set, I plane and sand the molding to final thickness.

—Bob Plath, Delhi, N.Y.

Shop-built doweling machine

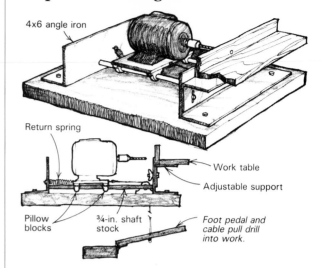

4x6 angle iron

Return spring. Work table. Adjustable support. Pillow blocks. ¾-in. shaft stock. Foot pedal and cable pull drill into work.

In my shop I used a doweling jig to drill holes for dowel joints in cabinet door frames and the like. Although I found this procedure too slow, I found horizontal boring machines too expensive. I built the machine shown above for about $260, which included $120 of machine-shop expense. The machine consists of two opposed 4x6 steel angle iron sections bridged by two ¾-in.-dia. steel shaft-stock rails. Boring is accomplished by a ½-HP, 1725-RPM motor that slides down the rails on pillow blocks. The sliding action is provided by a pulley, cable and foot pedal arrangement. A strong coil spring attached to the back of the motor base pulls the motor back out of the hole when the foot-pedal is released. The work table adjusts vertically to accommodate different stock thicknesses.

Since the rails must be prefectly parallel I had a machine shop drill the critical rail holes in the angle iron pieces. The machine shop also threaded the rail ends and reworked the motor shaft to accept a chuck. —Hjardar Brunn, Ferndale, Wash.

Making tight leg tenons

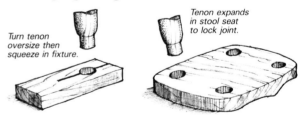

Turn tenon oversize then squeeze in fixture. Tenon expands in stool seat to lock joint.

I make Windsor stools with legs that have tenons turned to fit blind holes bored in the stool seat. These leg joints take a lot of stress, so I take pains to ensure the leg joint is tight. First, I turn the tenon slightly oversize, so its diameter is about ¹⁄₆₄ in. too big for the hole. Then, before assembly, I compress the tenon with the fixture shown in the sketch, so it fits into the hole in the seat. Later, because of the moisture in the glue, the leg will expand and lock itself in the socket.

The compression fixture is a maple board drilled with the same size hole as the tenon, then slotted halfway with a sawkerf. To compress the tenon, I insert it in the fixture and squeeze the fixture in a vise. I rotate the leg 90° and squeeze it. Scoring the tenon lengthwise will allow excess glue to escape.

—John Taylor, Golcar, Huddersfield, England

Sanding and Finishing

Chapter 8

Two-faced sanding slab

I suspect many among us like to sand small pieces of wood by rubbing them back and forth on a whole sheet of sandpaper, finger-pressed against the top of a workbench or the flat table of a handy woodworking machine. And just as many of us know that it's only a matter of time before… we slip and that fresh sheet of sandpaper is wrinkled or torn. If this sounds familiar, then between projects make this versatile sanding slab from a piece of scrap and a couple of inner-tube ribbons. The device firmly clamps a full sheet of sandpaper for sanding, but allows easy replacement when it's worn out. While you're at it, make two slabs so you can have four different grades of sandpaper at the ready, simply by flipping the slabs.

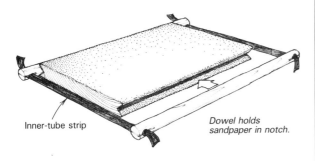

Inner-tube strip

Dowel holds sandpaper in notch.

Size the slab as long as a sheet of sandpaper but about an inch narrower so that you can fold the sandpaper's edges over into the V-grooves and hold them with the tensioned dowels. The thickness of the slab is not important. No doubt, ¾-in. stock and ¾-in. doweling would work just fine.

To use the slab, simply fold a sheet of sandpaper over its face, snap the tensioned dowels into the V-grooves and start sanding. Here's a hint for mounting two sheets at once: Tack two sheets of sandpaper in place temporarily with masking tape before snapping the dowels into the V-grooves.

—*Frank Schuch, La Mesa, Calif.*

Stripping with sawdust

To remove an old varnish finish quickly and neatly, first apply varnish remover and keep it wet until the finish has softened, then use handfuls of sawdust to remove the sludge. The sawdust acts as an abrasive and effectively cleans off the old finish. In addition, it absorbs most of the mess and makes cleanup an easy broom-and-dustpan task. —*Bill McNutt, Guthrie, Okla.*

Sandpaper tearing tool

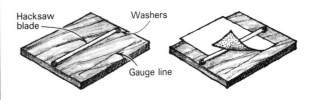

This idea has been around for a while, but this tool is one of the best I know of for cutting sandpaper sheets. Simply screw an old hacksaw blade to a scrap of plywood with washers under the blade for spacers. Mark standard sheet sizes on the plywood, or attach a rule for measuring sheets. To use, simply slide a sheet of sandpaper under the blade and pull for a quick, neat cut . —*Rick Mattos, Vallejo, Calif.*

Laminating sandpaper for flatness

Small items are always difficult to sand. No matter how careful you are, they never seem to come out flat. Try using rubber cement to hold the sheet of sandpaper to a scrap piece of laminated countertop from your local cabinet shop. Sink cutouts seem always to be in plentiful supply. The sandpaper will now be flat and you can use both hands to steady the piece being sanded. When the sandpaper is worn, simply peel it up and remove the rubber cement from the laminate by rubbing it with your finger. —*Robert A. Prive, Essex Junction, Vt.*

Sheet-metal sanding shield

Whenever I sand panel frames or other woodwork where it is difficult to avoid getting cross-grain scratches on adjacent surfaces, I use a very thin piece of sheet metal in much the same way as a draftsman uses an erasure shield. I hold or clamp the shield over the section I want to protect and then just sand right up to and over it. In a similar way I can drill or cut out shapes from the center of a sheet that allow a tenon or plug, for example, to stick through and be sanded without affecting adjacent areas. The sheet metal I use is some 28-gauge stainless steel that I found at a surplus and salvage store. It's thin (less than ¹⁄₆₄-in.), yet can withstand occasional belt sanding.

—*Sandor Nagyszalanczy, Santa Cruz, Calif.*

Melting shellac sticks with a hot-glue gun

The experts say that melting a shellac stick with a hot knife is the best way to fill imperfections (see *FWW #34*). But my lack of expertise with a hot knife produced an awful mess. So I retrieved my hot-glue gun from my box of things I wish I had never bought and discovered that the glue gun is an excellent shellac-stick applicator. It heats the material to just the right temperature and puts it just where I want it. To conserve material, I cut off only the amount of shellac I need and use a short length of dowel as a piston to push the shellac stick through the barrel of the gun.

—*Stephen Kelly, Birmingham, Ala.*

Rubber sanding block

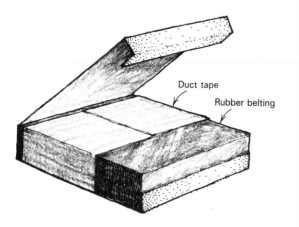

Duct tape

Rubber belting

The best solutions are always the simplest. When I needed a firm yet pliant sanding block for smoothing a long curve, I put together the block shown here, using two scraps of ⁵⁄₁₆-in. rubber belting and duct tape. Size the block so that ⅓ of a standard sheet of sandpaper will wrap around with about ½ in. left over on each end to insert between the two pieces of belting. The block's advantages include two fresh surfaces per filling and less sandpaper waste than commercial rubber sanding blocks. It is cheap and easy to make, and a snap to load. For a good fit, fold the sandpaper around the block and crease the corners before inserting the ends.

—*R .G. Sapolich, Johnstown, Pa.*

Two-faced sandpaper

Two-faced sandpaper, produced by sticking two pieces of sandpaper back to back with double-sided tape, is easier to work with because it doesn't slip under your fingers. The double-faced paper will also stick to a sanding block if the pad area is covered with flannel cloth. The paper won't slip on the flannel but it will pull off easily, so you can change grits almost instantly.

—*Joel B. Johnson, Hendersonville, N.C.*

Making contoured sanding blocks

When you use intricate molding in your work and insist on a perfect finish, the time invested in making a reverse-image, contoured sanding block is justified, even if you have to hand-carve the block. But when you can cast a perfectly accurate

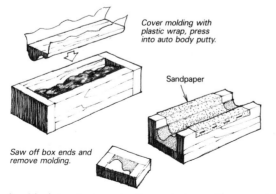

Cover molding with plastic wrap, press into auto body putty.

Sandpaper

Saw off box ends and remove molding.

sanding block in minutes, using the workpiece as its own mold, there's no excuse not to have one.

To make the sanding block, use scrap wood to construct a small box that's as wide as the molding and about 6 in. long. Mix up a small quantity of polyester auto body filler (I used Bondo) and partially fill the box with the putty. Now, cut a section of molding nearly as long as the box, cover it with thin plastic wrap and press the molding face-down into the body putty so that the air is expelled and the putty takes on the reverse shape of the molding. Hold the molding in place with C-clamps while the filler sets up. After the filler has hardened, bandsaw both ends of the box to free the molding and produce a U-shaped sanding block.

To complete the sanding block, tape sandpaper to a length of molding and sand the interior of the block until it is smooth. Then staple sandpaper to the sanding block, carefully folding the paper where necessary to fit small corners and narrow beads.

—*Earl J. Beck, Oak View, Calif.*

Foam sanding block

I don't know why, but we have a hard time keeping sanding blocks around our college shop. One evening, as time for class approached, I suddenly found myself out of them. In desperation, I quickly bandsawed a number of blocks from 1½-in.-thick plastic-foam insulating board, figuring they would last just long enough to get through that class. To my surprise, the fragile foam blocks held up well and proved to have unanticipated advantages. They were not only lighter than wood blocks, but they also conformed to uneven surfaces better than the felt blocks I had been using.

—*Mark White, Kodiak, Alaska*

Foam finish applicators

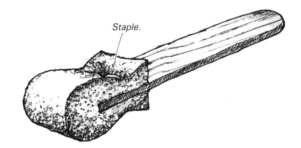

Staple.

You can easily make throwaway finish applicators from urethane foam, which is commonly sold in fabric and upholstery shops. Cut the foam to a 1x1x2 size, split it down the middle and staple it to a scrap of thin wood for a handle. You can then trim the free end with scissors to suit the job.

—*David E. Price, Baltimore, Md.*

Smoothing turned goods with cloth

Years ago I watched craftsmen at a Virginia shop put the final finish on lathe turnings by holding a piece of cloth against the work after sanding. Later I read that textile companies have started using ceramic yarn guides because synthetic yarns are abrasive and cut into steel guides quickly. Putting the two observations together, I tried finishing turned chair parts by first sanding with 000 garnet paper, then holding a scrap of synthetic-fiber drapery material against the spinning work. The cloth picked up grit left on the wood, and in less than a minute did indeed give the pieces a smoother look and feel.

—*Carlyle Lynch, Broadway, Va.*

Hiding hairline cracks in wood

Like many other craftsmen, I've been through the mill trying to find a suitable material for patching cracks, holes and other imperfections in wood projects. I finally hit upon a terrific solution: acrylic modeling paste—the kind artists use for thick, built-up effects. It's available at any well-stocked art-supply shop. You can color the paste to match any wood, using commonly available acrylic artist's paints. The paste will go into hairline cracks and can be piled up about ⅛ in. thick without cracking. It carves, sands and machines like wood. What's more, it will take any finish.

—*John Stockard, Milledgeville, Ga.*

Here's how to repair and fill a hairline crack that mars an otherwise usable piece of wood. You'll need fast-penetrating cyanoacrylate glue and extra-thick cyanoacrylate glue. Both are commonly available at hobby and model shops. If the crack is closed, hold it open with a knife. Apply the fast-penetrating glue first, which will be sucked deep into the crack by capillary action. Then apply the heavy-bodied glue, which will follow the thinner glue into the crack. Open and close the crack a few times to distribute the glue. If the crack is open, force wood dust into it with a spatula or an artist's palette knife and mix it with the glue. Clamp the wood if needed. Two hours is enough drying time.

—*John W. Wood, Tyler, Tex.*

Waterproofing turned vases

To prevent water damage to turned flower vases, I have tried built-up plastic finishes and even melted candle wax. Neither will last permanently and a failure will ultimately ruin a beautiful piece of wood. Glass test tubes, available at chemistry supply stores, provide the solution. They are available in a wide range of diameters and lengths to suit your needs. With a sharp spade bit, drill a hole in the vase slightly deeper than the test tube you will be using. There is a lip on the tube that will rest on the surface of the vase and allow for easy removal for cleaning later.

—*William D. Vick, Jr., Rockville, Va.*

Removing black water stains from oak

To bleach out black water stains on oak use a 20% solution of phosphoric acid. For safety's sake, don your goggles and rubber gloves, then just brush the acid solution on the oak and put it out in the sun. Neutralize the acid after it is dry with a TSP (trisodium phosphate) or bicarbonate of soda solution. I use this procedure on old oak barrels and find it more effective than the two-step oxalic-acid system sold in paint stores. The phosphoric acid also removes rust deposits from iron and steel, much the same as Naval Jelly.

—*Peter S. Birnbaum, Sebastopol, Calif.*

Low-cost airbrush

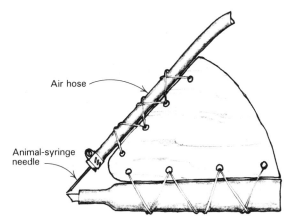

Air hose

Animal-syringe needle

You can make an effective airbrush by using a needle from a No. 11 animal syringe and a common felt-tip marker. As shown in the sketch, a simple wooden block with rubber bands holds the tip of the marker in the fine airstream that passes through the needle. For the air supply, use a shop compressor regulated at from 15 to 30 psi. I recommend the use of an electric solenoid to start and stop the air with a minimum of bleed-off. I use the airbrush to detail fishing lures in a rainbow of colors.

—*Fred J. Steffens, Monroe, Wis.*

Tablesaw Jigs and Fixtures

Chapter 9

Making fluted panels on the tablesaw

When I needed two 7-ft.-long fluted panels for the front entrance of my house, I made this simple angled fence and cut the flutes on my tablesaw. The fence is simply a wedge-shaped piece of plywood screwed to a plywood channel that press-fits on the rip fence. The angle of the plywood to the blade will determine the shape of the flutes, with a small angle producing deep, narrow flutes and a larger angle producing shallow, wide flutes. I found that an angle of 13° was about right for this particular job.

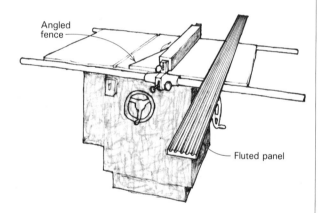

Angled fence

Fluted panel

To cut the fluted panels, fix the angled fence on the rip fence and set the blade depth at about ⅛ in. It will take four passes, raising the blade ⅛ in. each pass, to cut the flute to its final depth of ½ in. After one flute has been completed, move the rip fence over, lower the blade and start another.

—*Wayne A. Kulesza, Chicago, Ill.*

Angle finder for cutting coves

To cut coves accurately on the tablesaw, you must clamp the fence to the table at the correct angle, because the angle determines the shape of the cove. This simple parallelogram device makes finding the correct fence angle easy.

To use the device, simply move the arms apart a distance equal to the desired width of the cove and tighten the wing nuts to lock the arms at this distance. With the sawblade set as high

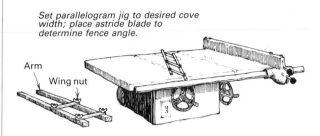

Set parallelogram jig to desired cove width; place astride blade to determine fence angle.

Arm

Wing nut

as the cove will be deep, place the device over the blade and rotate it until the arms contact the tips of the teeth at front and back. The device is now at the correct fence angle. All that remains is to clamp a fence to the saw at the same angle, and taking about 1/16 in. at a pass, begin making incremental cove cuts until the final depth is reached.

—*Joe Hardy, Des Moines, Iowa*

Making fluted panels revisited

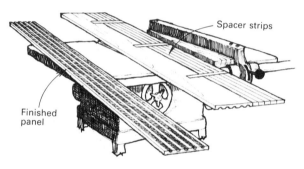

Spacer strips

Finished panel

Here's an improvement to Wayne Kulesza's method for making fluted panels (*FWW* #67, p. 8). First, determine the proper fence angle to produce the flute width desired, then clamp an auxiliary fence to the saw. Tape several spacer strips to the side of the workpiece; the width of the spacer strips determines the distance between flutes. Push the workpiece and spacers through the saw to cut the first flute. If your blade is sharp and the flute not too deep, it should be possible to cut each flute in one pass. After you have sawn the first flute, slice off the outermost spacer by cutting the tape with a utility knife. Continue making passes and removing spacers until you have completed the panel. The spacer strips eliminate the time required in resetting the fence for each pass through the blade.

If the blade binds and prevents you from cutting each flute in one pass, there are two alternatives: You can raise the blade in increments as you cut each flute, or you can make a series of shallow flutes across the width of the board, then replace the spacer strips, raise the blade and repeat the series until full depth is achieved. To ensure equal depth on all flutes, I would choose the second alternative.

—*Joe Videtic, Joliet, Ill.*

Panel-raising fixtures

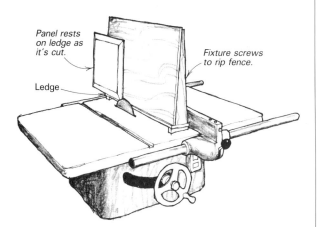

Panel rests on ledge as it's cut.

Fixture screws to rip fence.

Ledge

I made this fixture to expedite making dozens of ½-in.-thick, red oak panels 1 ft. wide by 5 ft. tall.

The height of the sawblade and the slope of the fixture produces panels beveled at 7½°, with the shoulder of the field cut simultaneously. If you push the panels through carefully and use a high-quality carbide blade, you'll produce panels virtually free of machine marks, requiring minimal sanding.
—*Warren W. Bender Jr., Medford, N.Y.*

Making twelve-sided turning blanks

It's common practice to rip the corners from square turning stock, making it octagonal, in order to reduce the amount of material to be roughed off in the lathe. It's just as easy to produce a twelve-sided turning blank, which requires even less stock removal in the lathe and doesn't require racking the saw blade over to 45°, where my tablesaw, at least, doesn't behave very well.

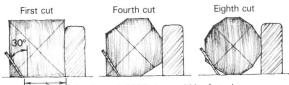

First cut   Fourth cut   Eighth cut

30°

Set fence 0.7887 times width of stock.

First, set the tablesaw blade at 30° away from vertical. Set the rip fence at 0.7887 times the stock width from the blade. This setting can be approximated accurately by taking ¾ in. plus 1/32 in. for each inch of face width. Mark both ends of the stock with an X, as shown in the sketch, so as to be able to keep track of which corners to cut off. Rip the corners from the square, turning the stock 90° clockwise each time if your fence is to the right of the blade. Now turn the stock end for end and repeat the process, keeping the original faces, which are now not the widest ones, against the fence and saw table.

A 10-in. saw will handle a 7½-in.-thick blank. The procedure is inadvisable for stock less than 1½ in. wide, because the faces become too narrow to guide safely against the fence and table.
—*Warren Miller, State College, Pa.*

Flexible hold-in

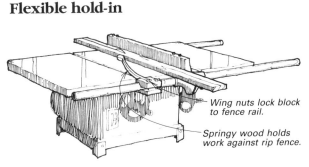

Wing nuts lock block to fence rail.

Springy wood holds work against rip fence.

Most featherboards or hold-ins utilize an angled board with numerous sawkerfs cut into one end. The flexibility of these featherboards is pretty limited, so they must invariably be reset after every cut. The alternative design shown in the drawing offers a much greater range of flexibility and requires fewer adjustments as ripping progresses. The bandsawn spring is dadoed into a split block that slides on, and locks to, the rip-fence rail for quick adjustment. The length of the spring and the strength of its action can be tailored to suit. Hickory or pecan are common springy woods that adapt well to this type of use.
—*Bert G. Whitchurch, Hemet, Calif.*

No-hassle octagon ripping

After spending a week making a marking gauge for laying out octagons on any size stock, I had a flash of inspiration. This method, which requires no gauge, will allow you to rip perfect octagonal cylinders.

First determine the size you want the finished octagon to be, then rip your stock to a perfect square. Make a new wooden insert for your tablesaw and, with the saw blade tilted to 45°, bring the blade up through the new insert to the maximum depth of cut. Retract the blade to the depth needed to cut the corners off the square stock. A precise kerf line should now be visible fore and aft of the blade in the tablesaw insert.

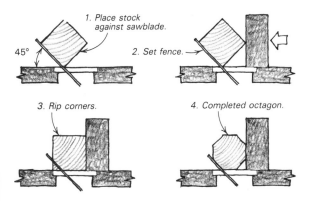

1. Place stock against sawblade.

45°

2. Set fence.

3. Rip corners.

4. Completed octagon.

Now lay your square stock against the blade with the corner of the stock right on the kerf line in the insert. Bring the rip fence up to the stock so that it just touches the corner (as shown in the sketch) and lock. Lay the stock flat on the table against the rip fence and rip off the corners to produce a perfect octagon.
—*L.A.D. Colvin, Satellite Beach, Fla.*

Octagon formulas and jig

The special jig and the formulas below will enable you to cut an octagon with each side equal to a predetermined length. The jig is a piece of plywood with two fences screwed to the top at 45° to the edge.

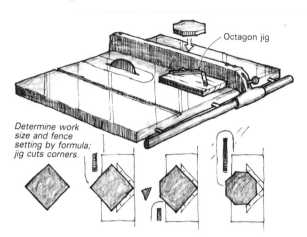

Octagon jig

Determine work size and fence setting by formula; jig cuts corners.

To use the jig, first determine the desired length of one side of the finished octagon O. Calculate the square size S needed from the formula $S = 2.414O$, and cut a square S inches on each side. Now calculate the rip fence distance R from the formula $R = 2.914O$, and set the rip fence at this distance. Place the square in the jig and rip off all four corners in turn to produce a perfect octagon.

Example: 3 = desired length of one side of octagon. $S = 2.414 \times 3$, or $S = 7.242$. $R = 2.914 \times 3$, or $R = 8.742$.

—*Rafik Eskandarian, Fresno, Calif.*

Self-clamping featherboard

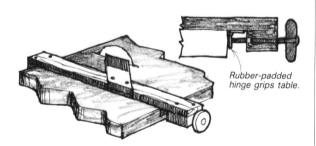

Rubber-padded hinge grips table.

Most woodworkers recognize the value of using a featherboard when feeding narrow stock through a saw or shaper. But too often we fail to use the device because it's simply too much trouble to clamp and adjust. This featherboard is no more trouble to adjust and use than a rip fence. It has paid for itself many times over in time and material savings. My version grips the table by means of a rubber-padded hinge, activated by a bolt running through a T-nut, as shown in the sketch—the handwheel is held on by epoxy. You could adapt the design to grip securely on virtually any rip-fence rails.

—*Bert Whitchurch, Rockaway Beach, Mo.*

Improved featherboard

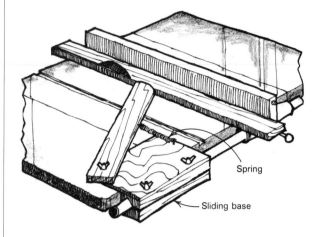

Spring

Sliding base

I finally got tired of the clumsy business of clamping a featherboard to the saw table, and then tediously reclamping it each time to adjust it to the width of a new workpiece. This simple solution took less than an hour to make and works perfectly.

It consists of two parts, a featherboard and a sliding base assembly. Custom-fit the sliding base to your front fence rail so that it can move anywhere along the front edge of the saw table and be locked in place with wingnuts or wedges. My sliding assembly is made to fit the T-slots of the Rockwell Unifence arrangement (the one in the drawing is shown on the more usual Unisaw rails). The featherboard pivots on a bolt and is kept in tension against the workpiece by a spring.

—*Arthur Kay, Tucson, Ariz.*

Featherboard variation

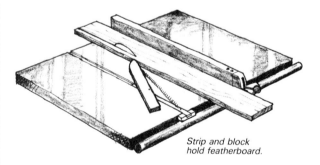

Strip and block hold featherboard.

I use a variation of Arthur Kay's featherboard (*FWW* #55) that is very quick to install and remove. My featherboard is fastened to a strip of 1/2-in.-thick wood that's a snug fit in the miter-gauge slot. A small block on the end prevents the strip from sliding toward the back of the saw. The featherboard itself is fastened to this strip from below by a screw and is held in tension against the stock being ripped by a spring. I reversed the traditional shape of the end of the featherboard so that it would bear on the stock as closely as possible to the front of the blade, yet still clear the teeth when the stock being ripped passes through.

—*Harold W. Books, North Platte, Neb.*

Foot switch for tablesaw

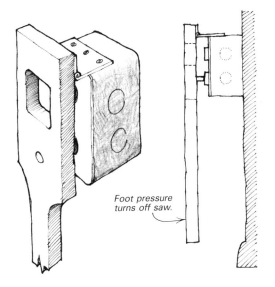

Foot pressure turns off saw.

This foot switch is for those of us, who, with both hands critically occupied on top of the saw table, have wished for a third hand to reach under the table and turn off the saw. The foot switch is simply a hinged paddle that hangs down over the saw's push-button switchbox. I can turn off the saw by bumping the paddle with knee or foot—a short dowel located at just the right spot pushes the off button. A hole through the top part allows normal finger access to the on button and, in fact, offers some protection against the button being pushed accidentally.

—*Eric Eschen, Chico, Calif.*

Remedy for a worn miter gauge

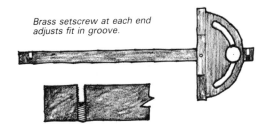

Brass setscrew at each end adjusts fit in groove.

Here's a more elegant way to take out the slop in a loose or worn miter gauge rail than peening the rail. First dismantle the gauge. Drill and tap a hole across each end of the rail bar to accept a ¼-in. setscrew. I make my own brass setscrews by cutting the head off a brass bolt and hacksawing a screw slot. Install the setscrews in the bar and adjust them until the rail fits the slot perfectly. —*Harrie E. Burnell, Newburyport, Mass.*

Lubricating tablesaw adjustment gears

To lubricate tight, binding adjusting gears in the tablesaw, first vacuum and then brush the mechanism with a nylon parts-cleaning brush. Then spray the gears with a chain lube such as Whitmore's Open Chain Lubricant or PJ-1 Heavy Duty Chain Lube. These slippery-film lubricants are well adapted to the dusty environment under the saw's table. An occasional application will provide continued smooth adjustment action, even when cutting abrasive materials like fiberboard or Masonite.

—*John Grew-Sheridan, San Francisco, Calif.*

Auxiliary tablesaw switch

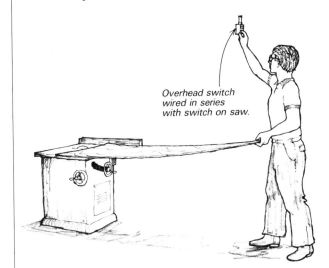

Overhead switch wired in series with switch on saw.

Here's how I solved the problem of operating the start/stop switch on my tablesaw when ripping large sheets of plywood. I installed a switch identical to the one on the saw at a convenient overhead location near the front of the saw. I wired this switch in series with the saw's switch so that both have to be "on" to let the saw run, but either switch will turn the saw off.

In use, I first make sure the overhead switch is off, and then I switch the saw on. I get the plywood into position and then reach up and flip on the auxiliary switch. For normal operations, I leave the overhead switch in the "on" position and use the saw's switch. —*Charles W. Leffert, Springfield, N.J.*

Magnetic duplicate cut-off aid

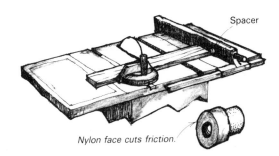

Spacer

Nylon face cuts friction.

This gadget has saved me a lot of time when cutting duplicate lengths on the table saw. It's a spacer that ensures clearance between the cut-off stock and the rip fence, thereby avoiding the danger of kickback. My spacer is simply a round magnet with a threaded hole through it. These magnets should be standard items at your local hardware store or five-and-dime. Screw a short length of nylon rod or other slippery alternative to the magnet and cut the unit to exactly 1 in. in length. I trimmed mine to length by facing it on the lathe, but other methods would work just as well.

To use, just pop it on the rip fence, set the fence to the desired dimension plus 1 in., butt your stock against it and cut. Best of all, the spacer is always handy—stuck on the back side of the fence.

—*Richard Bolmer, Anaheim, Calif.*

Holding push sticks with hook and loop fastener material

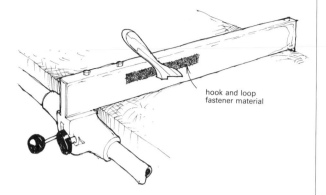

hook and loop
fastener material

Frustrated at never having my push sticks on my tablesaw when I need them, I glued a strip of hook and loop fastener material on the right side of my rip fence and the mating hook and loop fastener material on the sides of my push sticks. The push sticks now stand at attention on the rip fence, ready to be grabbed when needed. Most hook and loop fastener material now comes with a peel-off sticky back that should fasten the material well enough. This tip could be used for many accessories and tools around the shop.   —*David Crawford, Brownsboro, Tex.*

Improved push sticks

In the past eleven years I've worked in several shops, from a furniture factory to a custom cabinet shop, and taught some high-school woodworking as well. As you might imagine, I've seen my share of tablesaw push sticks, most of which resemble

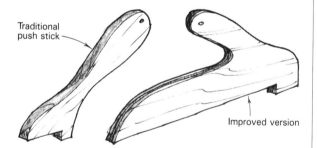

Traditional
push stick

Improved version

the design shown in the sketch. The major shortcoming of this design is that it does little to counteract the tablesaw's tendency to lift the work off the table.

The alternative design, shown on the right, ensures a downward thrust on the work. In production runs, this design is safer and less fatiguing because you need only push forward, rather than forward and down. You can make wide push sticks from solid wood, but for narrow ones choose plywood; otherwise the step on the bottom of the stick may split off.

—*Angelo Daluisio, Lancaster, N.Y.*

Cutting wedges on the tablesaw

I like to use the wedged through-tenon joint, and have developed a way to cut consistently tapered wedges quickly on the tablesaw. The key to the method is to save the waste cut-off

ends of glued-up tabletops and the like. I use a tablesaw jig to taper one edge of these scraps at ½-in. per foot, or 2° (the taper shown in the sketch is exaggerated to about 5° for clarity). To

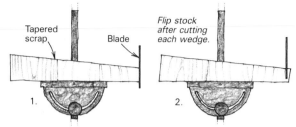

Tapered
scrap

Blade

Flip stock
after cutting
each wedge.

1.

2.

cut wedges I simply crosscut the tapered scrap piece, flipping the piece after each cut. Rather than measure, I eyeball the width of the wedge, making sure to cut the thin end of the wedge smaller than needed. When fitting the wedge, I trim off the thin end with tin snips until the wedge fits the kerf in the tenon perfectly.   —*C.M. Chappell, Houston, Tex.*

Roller support for crosscutting

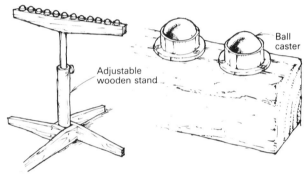

Adjustable
wooden stand

Ball
caster

Although there are many freestanding roller stands available for supporting work being ripped on the tablesaw, I've never seen a similar support for crosscutting. The support I developed has served me well in both ripping and crosscutting. Its principal component is a unique ball-shape caster (available from The Woodworkers' Store, 21801 Industrial Blvd., Rogers, Minn. 55374-9514). The caster rides on ball bearings, allowing smooth movement. Keep them clean and free from sawdust or they'll clog and won't run freely. The wooden stand is adjustable and has a weighted base.

—*William A. Lemke, Hendersonville, N.C.*

Roller-stand adjustment mechanism

The main differences between this roller stand and other designs I've seen are the ease of adjustment and the tilting head. The tilting head, together with the three-legged base, makes it easy to cope with a shop floor which, like mine, isn't level.

The arrangement for adjusting the height consists of a bolt threaded through a T-nut pressed into the inside of a cover plate, as shown in the drawing. The head of the bolt is captured in an oak knob, which permits easy hand-tightening. The bolt does not bear directly on the sliding dowel, but on a loose

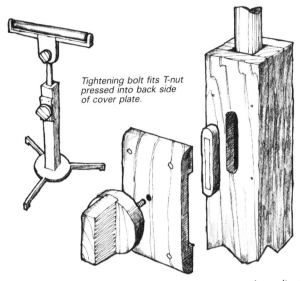

Tightening bolt fits T-nut pressed into back side of cover plate.

wooden insert set in the post. This permits smoother adjustments and prevents the bolt from chewing up the dowel. The insert itself is protected by a brass wear strip.

—*Timothy D. Anderson, St. Paul, Minn.*

Roller-stand adjustment revisited

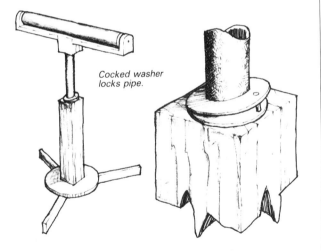

Cocked washer locks pipe.

My three-legged roller stands are similar to Tim Anderson's. But I have substituted a 1-in.-dia. steel pipe in place of Anderson's sliding dowel, and I use a cocked washer mechanism for adjustment and locking. It's the same idea found on many pipe and bar clamps. To make the adjustment mechanism, first locate a couple of steel washers with holes about ⅛ in. larger than the outside diameter of the pipe. Rivet a ½-in.-long steel bolt to one side of one washer. The bolt will cock the washer against the pipe and automatically lock the roller assembly at any height. The second washer prevents wear on the stand. To release the mechanism, simply lift up on the low side of the top washer.

—*Ingwald Wegenke, Montello, Wis.*

Plywood-scoring tablesaw insert

This device helps prevent the underside of veneered plywood from splintering by scoring the veneer just ahead of the blade. To make it, cut a snug-fitting insert blank from aluminum or plywood. Carefully raise the blade through it and enlarge the slot for clearance. With a straightedge against the teeth on each side of the blade, mark the outer edges of the blade's kerf on the front section of the insert. Cut along these lines with a thin saw to create two slots just ahead of the blade.

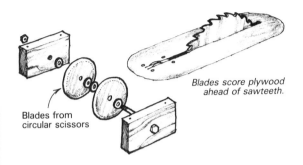

Blades score plywood ahead of sawteeth.

Blades from circular scissors

The scoring assembly uses two circular blades from the relatively new circular scissors made by Olfa. Spare blade packs for the scissors should be available at sewing and fabric stores. Make up a pillow-block assembly with a bolt axle to attach the scoring blades to the underside of the insert. Use regular washers, as well as shim washers punched from an aluminum can, to space the blades on the axle. Bolt the scoring assembly under the insert, making sure it will clear the sawblade and housing when in place. Ideally, the scorers should make a shallow cut (1/32 in. to 1/16 in.) in the plywood very slightly wider and in line with the kerf of the blade.

—*Sandor Nagyszalanczy, Santa Cruz, Calif.*

Tablesaw tenoning jig

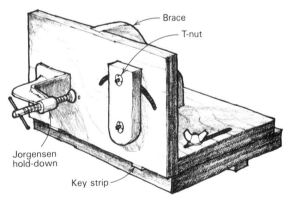

Brace

T-nut

Jorgensen hold-down

Key strip

Faced with cutting 32 tenons on a recent project, I priced a new Delta tenoning jig and found it beyond my budget. So I put myself to work and made the jig pictured here at the cost of a few ¾-in. plywood scraps and some hardware items already in my shop. It consists of a baseplate that rides along the miter-gauge slot, and a top table that adjusts closer to or farther from the sawblade. The work is clamped against both the jig's face and a vertical stop block, which can be pivoted to an angle if necessary. To cut the tenons, I installed two identical 8-in. blades on my saw arbor with a ½-in. spacer between. I had the 32 tenons cut in no time.

—*Harrie E. Burnell, Newburyport, Mass.*

Tablesaw rabbet/dado jig

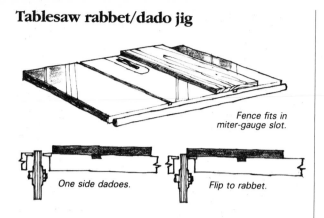

Fence fits in miter-gauge slot.

One side dadoes.

Flip to rabbet.

Because the plywood end panels I make for kitchen-cabinet jobs are usually worked the same way each time, I found myself setting the rip fence to the same measurements for dadoes and rabbets time and time again. The simple fixture above solved this problem because it is essentially a pre-measured rip fence that I can use instantly by just popping it into the miter-gauge slot. It's a dual-purpose fixture—I just lift it out of the slot, turn it end-for-end and push it back down into the slot to use the other side. One side cuts dadoes 2 in. from the edge of the workpiece, the other side cuts 3/4-in. rabbets.

The dado fixture worked so well that I made a second variation strictly for rabbeting. One edge is sized to cut 3/8-in. rabbets and the other to cut 1/4-in. rabbets. I discovered on this second jig that it's best to make the fixture to mount in the right-hand miter-gauge slot if your sawblade slides on the arbor from the right (and vice versa if your blade slides on from the left). If made this way, the fixture can be used with virtually any width dado head in the saw—any excess width in the dado head is covered and doesn't affect the rabbet.

—Don Russell, Auburn, Calif.

Sliding tablesaw carriage

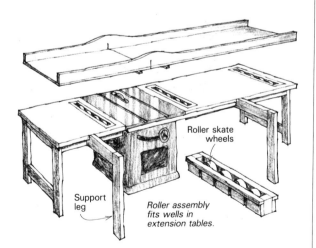

Roller skate wheels

Support leg

Roller assembly fits wells in extension tables.

If you've ever attempted to crosscut a 6-ft.-long, 2-ft.-wide panel on the tablesaw, you know the operation is awkward, error-prone and even scary. By contrast, when you add the sliding carriage described here, tablesaw crosscutting is made more accurate, faster and safer. The fixture is straightforward with two main components: an auxiliary bed fitted with rollers made

from skate wheels, and a sliding carriage that rolls atop the bed, using the miter-gauge slots as tracks.

The auxiliary bed fitted with rollers is really the key to the fixture. Without the rollers the heavy sliding carriage would stick and bind. To make the auxiliary bed, first construct two outrigger tables to bolt up to the saw as shown. The size of the tables is discretionary, of course, but I recommend that the tables-plus-saw add up to at least 8 ft. long. The rollers are made up in three box frames, each containing four nylon roller-skate wheels mounted on 1/4-in. threaded-rod axles. The roller boxes drop into wells in the top of each outrigger table as shown. The boxes pop out when they are not in use and can be replaced by plain plywood inserts. It is a good idea to design in some sort of height adjustment for the rollers in case they are too low or too high in use.

The sliding carriage is nothing more than a large panel of plywood with fences fixed to the front and back edges. Waxed maple runners screwed to the bottom of the carriage slide in the miter-gauge slots and ensure that the table tracks at right angles to the blade. I bolted the fences to the panel using slotted holes so I could adjust the fences for a perfectly square cut.

Because the sliding table is heavy, another necessary component of the fixture is a support stand to hold the table when it's pulled back toward the operator before the cut. You may choose to incorporate this support into the design of the auxiliary bed. In my case I made up a couple of removable legs I can fasten in place whenever I use the sliding table.

One side benefit of the fixture is that you may use the roller feature without the sliding carriage. The rollers make ripping a full sheet of plywood a breeze.

—Bill Amaya, Hailey, Ind.

Home for tablesaw accessories

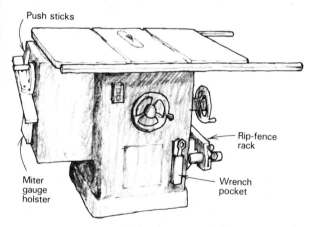

Push sticks

Rip-fence rack

Miter gauge holster

Wrench pocket

The last time I used my tablesaw on a project that required both crosscuts and rips, I couldn't find a place to park the rip fence and miter gauge to keep them safely close at hand. Also, the blade-changing wrench was always lost in the shop clutter, and my push sticks constantly wandered out of reach—just when I needed them most.

To resolve these problems, I decided to make homes for all my tablesaw attachments by building simple scrapwood holsters and racks at various places on the saw. The sketch above illustrates the idea.   —Fred H. Sides, Mt. Kisco, N.Y.

Safely removing small cutoffs

Place nozzle so it sucks up small cutoffs.

A good way to remove small cutoffs (such as chunks sliced off a dowel) from your tablesaw or bandsaw is to suck them in with your shop vacuum. Fit the vacuum's nozzle through a 2x4 notched to fit its hose diameter. Clamp this setup on the tabletop with the nozzle mounted as close to the cut-off point as possible. When you're done, the parts are neatly collected in the barrel. —*David Shaffer, Grand Rapids, Mich.*

Cutting small parts on the tablesaw

I made this fixture to cut parts for small models and miniature furniture. It works so accurately and safely that I cut even conventional-size parts with it instead of using the miter gauge.

The fixture is made from a piece of plywood that is 8 in. to 14 in. longer than the table, depending on the length of cuts you plan to make. Screw hardwood runners underneath the plywood to slide in the miter-gauge slots, and screw stop blocks at both ends of the underside to prevent accidentally cutting the fixture in two. With the fixture in place, raise the blade through it to cut the blade slot.

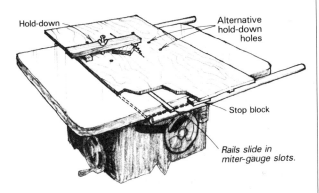

Hold-down
Alternative hold-down holes
Stop block
Rails slide in miter-gauge slots.

Drill several hold-down anchor holes through the plywood and install ⅜-in. T-nuts underneath. I have various anchor locations on my fixture to suit my individual operations. A small wood scrap will serve as a hold-down. Bore a hole through it for a ⅜-in. bolt, and thread a wing nut over the bolt before putting it through the wood block. Place the workpiece under the hold-down near the blade and place a block the same thickness as the workpiece at the other end of the hold-down—just tighten down on the wing nut to hold the workpiece in place.

I've found the hold-down applies enough force to lock the workpiece in place in just about any cutting operation, and my hands never come near the moving blade.

—*Don H. Anderson, Sequim, Wash.*

Two rip-fence straddling jigs

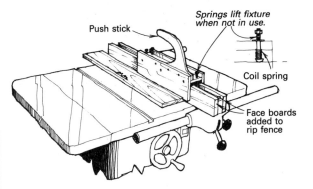

Push stick
Springs lift fixture when not in use.
Coil spring
Face boards added to rip fence

This fence-straddling push stick was originally designed to fit a Biesemeyer fence, but it could be adapted to practically any fence fitted with auxiliary face boards, as shown in the sketch. For added convenience, I installed plungers made from coil springs and bolts to raise the push stick when not in use.

—*Bill Hatch, Greensboro, N.C.*

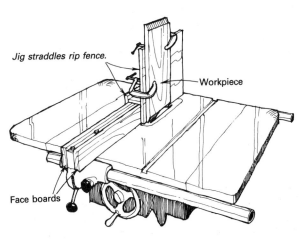

Jig straddles rip fence.
Workpiece
Face boards

This jig is designed specifically for making steep angled cuts on the edges of long workpieces, as in making fielded panel doors. If you've tried sliding a wobbling workpiece vertically along the fence, while watching to make sure the tapered end doesn't fall through the space in the saw insert, you'll immediately recognize the advantages of this jig. With it, you can make a smooth, controlled, burn-free cut.

Dimensions aren't critical, just make sure the jig slides smoothly on the rip fence. The face of the jig can be large or small depending on the size of the workpiece. It can be fitted with a vertical fence if needed. Just C-clamp the workpiece to the face of the jig, then slide the jig past the blade.

—*Alfred W. Swett, Portland, Maine*

Consistent dadoes on the tablesaw

The secret of this simple wooden tablesaw insert is the hump in the middle. Because the work will touch only at the high point of the hump, dadoes and grooves will be consistently the same depth regardless of slight waves and warps in the wood.
—*David Ward, Loveland, Colo.*

Flip-down wheels

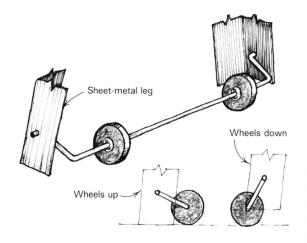

Sheet-metal leg

Wheels down

Wheels up

This simple flip-down axle fits tool stands with bent sheet-metal legs. First slide two wheels on a ⅜-in steel rod, adding washers and cotter pins to keep the wheels in position. Bend the rod to a wide U-shape, as shown, and install the axle through holes in two legs. The holes should be located so that when you lift up the end of the stand and flip down the axle with your foot the axle will bear against the inside bend of the leg, effectively locking itself in position.
—*Jeff Lormans, Dunedin, New Zealand*

Safe molding on the tablesaw

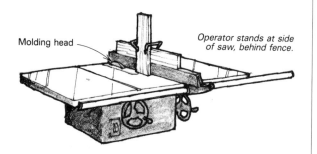

Molding head

Operator stands at side of saw, behind fence.

After I had a $1300 accident at my jointer last year, I have a renewed interest in safety. The scariest operation I know is using the molding head on the tablesaw to shape short vertical boards such as drawer fronts. I have rendered this operation relatively harmless by clamping the drawer front to a long board that rides the top of the rip fence. I guide the work through the blade, standing to one side of the saw behind the rip fence.   —*Richard Tolzman, Excelsior, Minn.*

Edging plywood drawer fronts

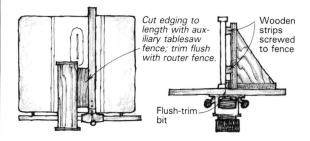

Cut edging to length with auxiliary tablesaw fence; trim flush with router fence.

Wooden strips screwed to fence

Flush-trim bit

Here are a couple of tricks I use to apply solid-wood edging to plywood drawer fronts. The first is a simple auxiliary tablesaw fence to trim the edging to length. I glue the edging to the ends of the drawer, leaving a ¼-in. overhang. Then, with the auxiliary fence adjusted for a perfect flush cut, I simply push each corner through the saw.

To trim the edging flush with the face of the drawer front, I use the router-table setup shown in the sketch. Make a tall fence for the router table and screw a couple of wooden strips to it. Chuck a ball-bearing flush-trim bit in the router and adjust the fence so the bearing is flush with the surface of the strips. When you run the panels through, the edging rides under the bottom strip and the tall fence makes it easy to keep the panel perpendicular.
—*Rick Turner, Petaluma, Calif.*

Precise tablesaw jointer disc

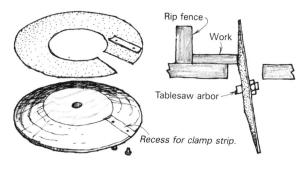

Rip fence

Work

Tablesaw arbor

Recess for clamp strip.

This cone-shaped tablesaw jointing device does precise work both in sanding boards to width and in leaving a good gluing surface. It's basically a 10-in. conical disc made from ¾-in. plywood. What makes the device special is the small cone angle (9° on mine). Unlike a flat disc, which contacts the workpiece across its full width, the cone's contact is restricted to a small area, the radius that's located directly above the arbor. Another benefit is that vertical adjustments of the saw arbor produce very small increments in cutting depth (as an industrial modelmaker, I sometimes have to work in thousandths of an inch).

I fiberglassed the back of my disc for extra strength, and it's served me well for years. The cutting surface is an abrasive sander disc cut to fit the cone face and glued and clamped in place. I use plastic-laminate glue to secure the paper; to change abrasives I dissolve the cement with acetone. Varying the grade of abrasive paper allows various compromises between a fast cutting speed and good surface quality.

To use the sander, tilt the tablesaw's arbor to present a vertical cutting surface, as shown. I usually guide the work along the rip fence. —*Dr. Robert Bogle, La Jolla, Calif.*

Tablesaw guard

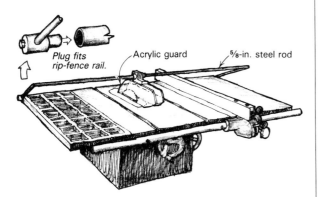

Unlike most other tablesaw-blade guards, which after a couple of frustrating experiences are left hanging on the wall, this guard is quite usable. The guard's main advantage is that it remains in place for most operations, including dado and molding cuts. If it's not needed for an operation, the guard swings out of the way in seconds, or can easily be removed completely from the saw. The inexpensive guard also acts as a hold-down—a safety bonus.

Make the guard shield from ¼-in. thick clear acrylic. The guard frame is a length of ⅝-in. cold-rolled steel bent into a U shape. Turn two metal or wooden plugs and attach them to the arms of the frame as shown in the sketch. The plugs should be sized to pivot easily in the holes in the ends of the back rip-fence rail. The frame fits on the saw by springing slightly so that the plugs snap into the holes.
—*K.L. Steuart, Ladysmith, B.C.*

Stationary jig for cutting open mortises

This jig is used to cut open bridle-joint mortises in frame members. It solves many of the problems inherent with sliding jigs, which tend to be complicated to make and adjust, and sometimes wobble during the cut. The only disadvantage is that the jig leaves a slight concavity at the bottom of the mortise, as shown in the sketch. This space doesn't show in the finished joint, however, and since it's end grain, the missing wood isn't critical to joint strength.

Make the jig by screwing a hardwood fence to an 8-in.-wide piece of ¾-in. plywood. Clamp the jig to the rip fence so that the frame member to be mortised will be centered over the saw arbor. Adjust the rip fence so that the sawblade is the proper cheek thickness from the jig.

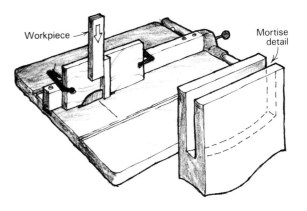

To cut the mortise, hold the workpiece firmly against the plywood, with its back edge tight to the hardwood fence. Plunge the work down the fence onto the blade. Draw it up, flip it and cut the other cheek. On narrow stock, this will complete the mortise. For wider stock, chisel out the waste.
—*Frederick J. Miller, Chatsworth, Ont.*

Powering other tools with a tablesaw

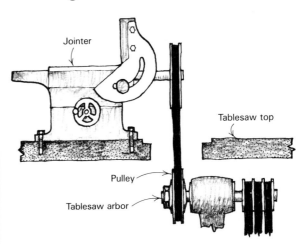

If you work in a small shop as I do and space is at a premium, here's how to drive another tool, a small jointer for example, with your tablesaw's motor. First slip a pulley on the saw's arbor and tighten the arbor nut as you would with a sawblade. As with any tool setup, be sure to size the pulley to drive your tool at the proper speed. Place the jointer on the saw table and clamp or bolt it in place. If you plan to use the arrangement often, you may wish to drill and tap fastening holes for the jointer right into the saw table. Next, slip a belt over the pulleys and lower the saw motor until the belt is tensioned. The whole arrangement takes no longer to set up than any other tablesaw accessory, and you save the price of a motor and gain a little space in your shop as well. —*Luc Mercier, Laval, Que.*

Slot mortiser

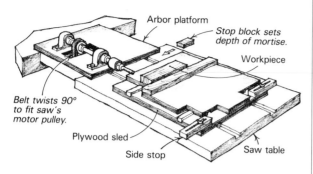

Arbor platform

Stop block sets
depth of mortise.

Workpiece

Belt twists 90°
to fit saw's
motor pulley.

Plywood sled

Side stop

Saw table

This low-cost slot mortiser utilizes my Sears saw table and motor. The arbor platform sits snugly in a space between the saw table and a catching table I built behind the saw. Short rails under the platform fit the miter-gauge slots to keep the platform from shifting. The V-belt on the arbor twists 90° and slips over the saw's motor pulley (the direction of twist determines which way the arbor will turn), and then the weight of the motor keeps the belt taut.

The sliding table consists of two plywood sleds that allow forward-and-back and side-to-side motion. Hardwood runners are attached to the lower sled, top and bottom. The top sled has two sets of grooves, so it can be repositioned atop the lower sled for end-grain mortising, at 90° from the position shown in the sketch. Various stop blocks limit the mortiser's travel and control mortise depth. To adjust for mortising stock of various thicknesses, shim the work or the arbor platform.

—Joel Katzowitz, Marietta, Ga.

Rip-fence extensions, two ways

My decision to extend the rip-fence rails on my Rockwell contractor's tablesaw came after a 4x8 sheet of plywood I was cutting "freehand" kicked back on me. I removed the tubular rails, took them to a machinist, and asked him to extend them so I could easily set the fence up to 48 in. The machinist welded a 2-in. steel plug in the end of each rail, drilled a ¾-in. hole through the plug, then tapped the hole with a coarse thread. He made the 24-in. extensions from tubular steel the same size as the rails and fitted each extension with a thread that screws into the plug in the original rails. These extensions have saved me countless hours of production time

—Stephen Seitz, Oleyo, N.Y.

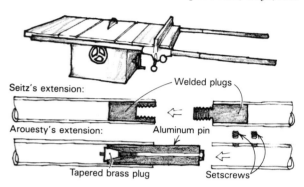

Seitz's extension:

Welded plugs

Arouesty's extension:

Aluminum pin

Tapered brass plug

Setscrews

To extend the rails on my Rockwell tablesaw, I purchased a second set of tubular rails and devised expanding aluminum pins to attach them to the original rails. I drilled a bolt hole through the length of each pin and slotted the inner end, as shown in the sketch, so it could expand to lock the pin in the rail. With the pins locked into the original rails, I slip the spare rails on and fasten them in place with setscrews, which are located on the inside so they'll clear the rip fence.

I'm pleased with this system because I can remove the rails when they're not in use and I can use the saw's racking "micro-adjustment" mechanism all the way across.

—Raymond Arouesty, Reseda, Calif.

Drill Press Jigs and Fixtures

Chapter 10

Cutting rosettes with a fly-cutter

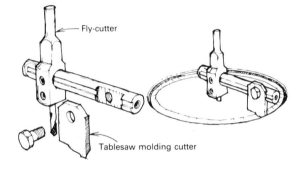

Here's how to convert a common drill-press fly-cutter to cut interesting circular patterns, or "rosettes," using an interchangeable blade borrowed from a tablesaw molding head. First, remove the swing arm from the cutter and file or grind a slot in the arm to fit the molding cutter. Size the width of the slot carefully so the cutter is snug and can't shift during use. Drill and tap a hole in the arm, and secure the molding cutter to the arm with a short bolt.

The unit works best with the fly-cutter's center drill bit acting as a pivot. You may want to replace the center bit with a tapered rod to leave a smaller center hole. If you must work without a center pivot, clamp the workpiece firmly to the drill-press table, run the drill press at its slowest speed and be careful.
    —*Donald F. Kinnaman, Phoenix, Ariz.*

Homemade bit for deep holes

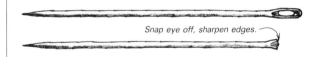

Hammer end of rod flat and sharpen.

To drill holes for long threaded rods, I hammered one end of a 26-in. steel rod flat and sharpened it as shown in the sketch. The bit won't pull chips out of the hole like an expensive ship's auger, so you'll have to retract it more often to clear the chips. Considering the savings, this is a minor inconvenience.
    —*Ralph Zwiesler, Freesoil, Mich.*

Cutting glass circles on the drill press

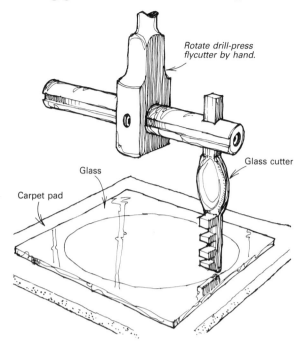

You can cut circles from glass or mirror using a drill-press fly-cutter and a modified glass-cutting tool. First, remove the cutting bit from your fly-cutter and replace it with a glass cutter that has its handle cut down and ground or filed to fit the hole in the fly-cutter. Adjust the device for the desired diameter circle and chuck it into the drill press. Place the glass on the drill-press table, with a thin carpet pad beneath the glass to absorb the shock. Lubricate the cutting wheel with kerosene, lower it onto the glass and lock the spindle so the cutter exerts light pressure against the glass. Then turn the drill press one revolution by hand. Caution: Don't do this under power; it would be dangerous. Besides, once around will do a better job.

Remove the glass from the drill press and make radial cuts with a glass cutter from the circle to the edge of the glass to help break the circle free.
    —*Bill Kilmain, Orlando, Fla.*

Making tiny drill bits

Drill bits of $1/32$ in. or less are hard to find, expensive and break easily. But in minutes you can make one from an ordinary sewing needle. These are readily and inexpensively available in many small sizes. To make a bit, use two pairs of pliers to snap the needle right at the bottom of the eye. The resulting blank is too hard for filing but, if held in a pin vise, can easily be stoned by hand to yield good cutting edges. (A Foredom Micro Chuck that will adapt your regular chucks to hold hair-thin bits is available from Woodcraft Supply, 41 Atlantic Ave., Woburn, Mass. 01888.) I use these needle bits not only in marquetry, where they are indispensable, but also for drilling $1/50$-in.-dia. holes $1/2$ in. deep in oak.
    —*Edward C. Kampe, Zellwood, Fla.*

Countersinking in cramped quarters

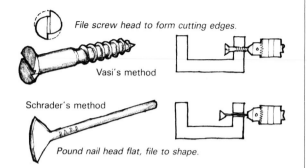

File screw head to form cutting edges.

Vasi's method

Schrader's method

Pound nail head flat, file to shape.

Here's a way to countersink a screw hole on the inside of a small drawer or other cramped space that you can't get a drill into. File both sides of a screw head to produce cutting edges, as shown in the top sketch, above. Insert the screw in the hole from the inside and chuck the screw's threads in a portable drill. A few revolutions will produce a perfect countersink.

—*James Vasi, Williamsville, N.Y.*

To countersink holes for wood screws when you don't have enough clearance for a drill, flatten and spread the head on a 10-d common nail. File the flat to the proper angle and cut the point off the nail. Poke the nail through the hole from the inside and chuck the shaft in an electric drill.

—*Robert H. Schrader, Carrollton, Ohio*

Doweling guide

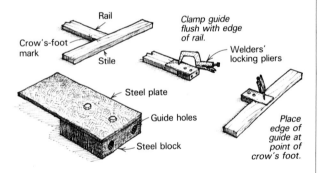

Rail

Crow's-foot mark

Stile

Clamp guide flush with edge of rail.

Welders' locking pliers

Steel plate

Guide holes

Steel block

Place edge of guide at point of crow's foot.

Before we discovered this simple steel guide for doweling the joints on cabinet fronts, we tried several other approaches, including an expensive two-spindle horizontal drill.

Make the guide out of a 1¾-in. square, ¾-in.-thick steel block. Drill two 5/16-in. holes (for 5/16-in. dowels) through the block. The holes should be spaced 5/16 in. in from each side of the block. The usability of the guide depends on the accuracy of these holes, so drill them precisely with a drill press. Complete the guide by bolting a 3½-in. length of steel plate flush to at least one face.

After your rails and stiles are cut, mark the joints as shown in the sketch. Clamp the guide to the side of the stile with one edge of the guide flush with the mark and drill the two holes. Then, use the guide to bore the two holes in the end of the rail.

Whatever marking system you use, just make sure you're consistent throughout. The drawing shows the crow's-foot

mark. The advantage of this mark is that if you place the guide so you can't see the crow's foot, then you're boring on the wrong side of the mark. —*Tim Hanson, Indianapolis, Ind.*

Disposable doweling jig

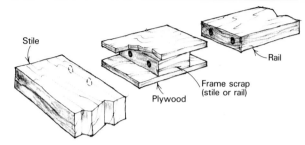

Stile

Rail

Frame scrap (stile or rail)

Plywood

When joining face frames, make this doweling jig from 1½-in.-long scraps from the frame's rail or stile and a piece of plywood. Bore the two guide holes in the block on a drill press, then glue the block between two pieces of ¼-in. plywood that extend 1 in. from each end of the block. Mark one face and one edge of the jig as reference surfaces to ensure consistency when drilling dowel holes. —*Ronald F. Seto, San Rafael, Calif.*

Shopmade screw-pilot drill

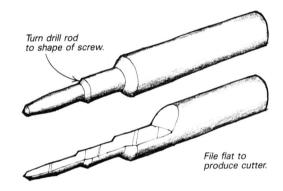

Turn drill rod to shape of screw.

File flat to produce cutter.

Most woodworkers are familiar with the special bits that drill a combined pilot hole, counterbore and countersink for screws. They are quite useful when you have a large number of screws to install. Unfortunately, these tools are not available for screws smaller than No. 6 or for other odd sizes and shapes. Here's how to customize your own to fit any screw.

Start with a length of oil-hardening drill rod, available at any industrial distributor or machine shop. Chuck the rod into a metalturning lathe and machine the end to the desired shape of the screw hole. If you don't have access to a metal lathe, chuck the rod in a drill press, a wood lathe, or even your electric drill and file the rod to shape while it is rotating. Taper the transition points to avoid sharp internal corners that would lead to stress points and possible breakage later.

To produce a cutting edge on the tool, file away exactly half of the cutter, taking care not to round over the edges. Now harden and temper the cutter with a propane torch (see *FWW* #50 for more on hardening and tempering). Whet the flat side with an oilstone and the cutter is ready to use.

—*John G. Martin, Cumberland, Maine*

Quick-change countersink

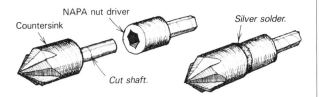

Countersink · NAPA nut driver · Silver solder.
Cut shaft.

I use a lot of countersunk drywall screws in my cabinet work, but I couldn't find a countersink with a six-sided shaft to fit my magnetic bit holder. As a result, I wasted a lot of time chucking back and forth between countersink and driver.

To solve this problem, I made a fast-change countersink by silver-soldering the shortened shaft of a common countersink into a NAPA nut driver. The driver's shaft easily slips into the magnetic bit holder.

—*Harry Sommers, Coeur d'Alene, Idaho*

Two hole-enlarging methods

To enlarge a hole in wood when you don't have the exact size bit you need, use a fly cutter with a hardwood plug over the pilot bit as a guide. Make a plug the same size as the hole to be enlarged, and wax the plug so it turns freely. If the hole is going to be enlarged only fractionally, cut a slot in the plug's side to hold the cutter bit, as shown below left. —*C. Dean Hawley, Okla.*

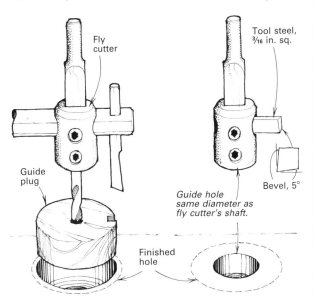

Fly cutter · Tool steel, ³⁄₁₆ in. sq. · Guide plug · Guide hole same diameter as fly cutter's shaft. · Bevel, 5° · Finished hole

When I needed to drill several precise 2-in. holes and my fly cutter proved unsatisfactory, I thought of replacing the crossbar in the fly cutter with a shopmade cutter, as shown above on the right. I ground a 5° bevel on a length of ³⁄₁₆-in. tool steel and locked it in place. Then I drilled a pilot hole the same size as the diameter of the fly cutter's shaft. This device bores clean, sharp holes that are amazingly accurate.

—*Samuel W. Pool, Cupertino, Calif.*

Drill-press cabinet

I use a standard stainless-steel hose clamp to attach a small cabinet to the post of my drill press. The cabinet, which holds drill bits and fixtures, hangs from a sturdy 1-in.-thick maple arm that has a shallow V-groove where it bears against the post. A

slot cut in the arm receives the hose clamp's strap to hold it to the post. I round off the edge of the slot nearest the post to keep the strap from crimping too much when tightened. This mounting scheme is surprisingly rigid, and the cabinet can be removed quickly or repositioned with just a screwdriver.

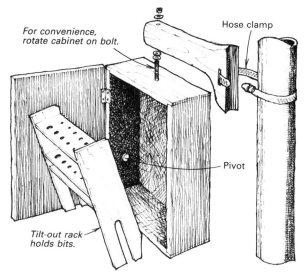

For convenience, rotate cabinet on bolt. · Hose clamp · Pivot · Tilt-out rack holds bits.

The cabinet is a simple box about 15 in. high, 8 in. wide and 3½ in. deep, and it is attached to the arm with a ³⁄₈-in. bolt. I put a couple of washers between the arm and the box so the box can be rotated to a convenient angle. The box is fitted with a standard drill-bit index on the bottom shelf to hold small bits and a shopmade rack above to hold larger bits. This rack flips forward so the bits can be removed easily without hitting the cabinet top. —*James H. Smith, Champaign, Ill.*

Modifying drill bits for brass

Here's a metalworking tip that is perhaps not common knowledge among woodworkers. When drilling soft metals such as brass, always grind or stone a small flat on the bit's cutting edge. This flat prevents the drill from chattering and results in much cleaner drilling.

—*Thomas J. Tidd, Springfield, Pa.*

Flat

Temporary micro-chuck

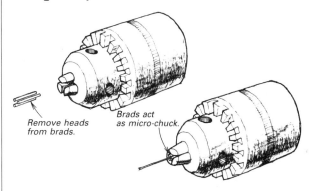

Remove heads from brads. · Brads act as micro-chuck.

When I needed to drill several tiny holes but didn't have a micro-chuck for holding a tiny drill bit, I used three brads as a micro-collet. I first clipped the heads off the brads, then slipped them and the bit into the chuck as shown in the sketch.

—*Paul Schulman, Belle Harbor, N.Y.*

Two drill-press sharpening systems

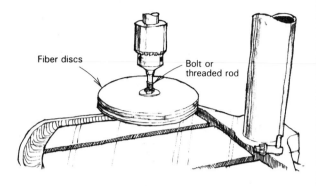

Fiber discs

Bolt or
threaded rod

This drill-press stropping wheel cost me about $2 to make and is handy to use because it spins horizontally and allows me to take advantage of the machine's low speed setting. To make the wheel, cut up four 6-in.-dia. circles of ¼-in.-thick cardboard or fiberboard. Stack the discs, drill a hole through their center and use a ½-in.-dia. bolt or threaded rod as an arbor, with washers on both faces of the wheel. Mount the wheel in the drill press, true the rim of the wheel and remove any projecting fuzz with coarse sandpaper. With a pocket knife, score a spiral groove on the wheel's face from a point near the arbor to the edge to catch honing compound. With the wheel spinning, apply rouge or tripoli to its face and edge.

To sharpen a chisel, I buff its flat side first on the wheel's face, with the wheel's rotation running away from the cutting edge. I hold the chisel at an angle to the wheel, and for greater control, work near the center of the wheel where its speed is slowest. I sharpen the edge of the chisel on the edge of the wheel, once again, with the wheel running away from the cutting edge. Two or three passes on the chisel's edge leaves it beautifully polished and razor sharp.   —*Leslie H. Blair, Rocky River, Ohio*

With my interests in furniture building, turning and, lately, chip carving, I always have many chisels to sharpen. This simple system helps me sharpen my tools quickly. The system consists of a 9-in.-dia. aluminum disc with an arbor made from a bolt or threaded rod and chucked in my drill press. I glue a circle of 200-grit wet-or-dry abrasive paper to the disc with Glop, an adhesive sold at many auto-parts stores. Rubber cement also works but breaks down sooner. I adjust the drill-press table so the disc spins about an inch or two above the water level in a plastic wash basin that contains an inch of water.

To sharpen a chisel, I set the drill at 300 RPM and wet the wheel's face with a squirt from a large bulb-shape syringe and then lay the bevel on the wheel's face. I sharpen the chisel for a bit and then squirt on some more water. It works so well that even sharpening my lathe tools on it has become a pleasure.
     —*Robert D. Panza, Canoga Park, Calif.*

Drilling centered holes in dowels

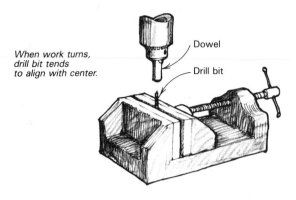

When work turns, drill bit tends to align with center.

Dowel

Drill bit

When you need to drill a longitudinal hole in a small dowel such as Jeris Chamey's box hinge (*FWW* #62), try this. Chuck the dowel in the drill press and hold the drill bit in the drill-press vise. When you lower the dowel on the bit, it will self-center and provide a quite accurately centered hole.

If your drill-press vise doesn't have a vertical slot milled in the jaws, here's how to align the bit. First, tighten the fluted end of the bit in the chuck with just enough pressure to hold it without damaging the flutes. Then, grip the shank of the bit with the vise. Release the chuck and the drill will be vertically aligned, ready to drill the workpiece.   —*Bob Grove, Portland, Ore.*

Rounding tenons on door louvers

Notch

Louver

Recently I had to make louvered shutters out of walnut for a custom interior job. Everything was straightforward until I got to the little round tenons on each end of about 500 louvers. When I cut the louvers, I left a ¼-in. square stub on each end, which needed to be rounded to pivot in the door frame. Stumped, I finally discovered that I could chuck a ¼-in. hex socket in my drill press and it would effectively chew the tenons round. The socket worked fine without any modification, but I had so many louvers to do that I decided to make the cutting action cleaner—I ground a notch in the socket, as shown in the sketch, to form a cutting edge. The edge should be ground at the middle of one of the socket's flats, at the point of minimum diameter.
     —*Paul G. Carson, Cashiers, N.C.*

Drilling accurate holes in large panels

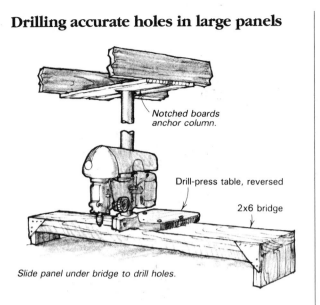

Notched boards anchor column.

Drill-press table, reversed

2x6 bridge

Slide panel under bridge to drill holes.

This idea evolved after I contracted to drill a series of precise holes in a pile of large panels. To support the drill press over the panels I built a 2x6 bridge as shown, and pipe-clamped it to a sturdy workbench. I removed the base of the drill press and used the drill table (reversed and rotated 180°) as an anchor to the bridge. I secured the top of the drill column by installing two V-notched boards on my shop's ceiling joists. Fences and stops fastened to the workbench position the panels for accurately spaced holes.　　—John D. Todd, No. Falmouth, Mass.

Lathe Jigs and Fixtures

Chapter 11

Turning tiny spheres

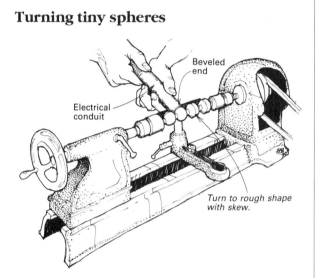

Beveled end

Electrical conduit

Turn to rough shape with skew.

The next time you need several small spheres for a project, try this technique. On a cylinder turned slightly oversize, lay out the number of balls you want, and with a parting chisel, make a cut to separate each ball segment. Be sure to leave at least a 3/8-in.-dia. section between the segments so the workpiece won't break.

With a skew, turn the balls to a rough spherical shape. Now take a section of thin-wall electrical conduit with a sharpened end and slip it over each rough sphere as it turns in the lathe. The conduit will cut the sphere to final size and give you a perfectly round ball. Be careful not to push the conduit too hard, or you will cut through the wood separating each ball.

—Donald F. Kinnaman, Phoenix, Ariz.

Pipe handles for bowl gouges

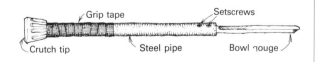

Grip tape   Setscrews

Crutch tip   Steel pipe   Bowl gouge

A steel pipe makes a nice handle for high-speed steel bowl gouges. It is not only heavier and more stable than a wooden handle, but it also allows the gouge to be adjusted in or out, depending on the application. Drill and tap two holes for setscrews near the end of the pipe for adjusting and securing

the gouge. To improve the grip, wrap the last few inches with tennis-racquet grip tape and put a rubber crutch pad on the end of the pipe.

—Earl R. Rice, Augusta, Ga.

Auxiliary lathe tool rest

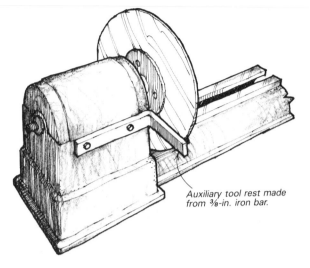

Auxiliary tool rest made from 3/8-in. iron bar.

My lathe has a 15-in. capacity over the gap, but when I mount a large bowl or tray, it's impossible to get the tool rest behind the blank to turn the bottom. To provide a tool rest for working the back, I bent a strip of 3/8-in.-thick iron to a 90° angle and bolted it to my headstock casting, as shown in the sketch. The tool rest was so useful that I made a set of them bent to different angles to fit different shapes. For safety's sake, remove the tool rest before sanding so your fingers don't get pinched.

—Kevin G. Weir, Brantford, Ont.

Sliding motor mount

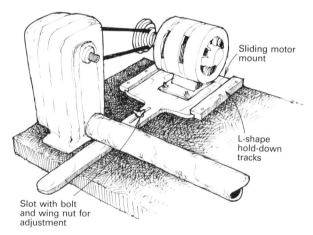

Sliding motor mount

L-shape hold-down tracks

Slot with bolt and wing nut for adjustment

This sliding motor mount for the lathe offers several advantages over the usual hinge-type mount. The device prevents the motor from jumping when a lathe tool gets hung up in the work, and it enables belt tension to be adjusted and remain constant when set. The mount consists of a plywood paddle on which the motor is attached and a couple of L-shape hold-down tracks. A slot in the paddle arm allows the mount to be locked at any position with a wing nut or to be loosened by hand for moving the belt to a different pulley.

—*Charles W. Whitney, Mount Vernon, Ohio*

Screw-drive centers

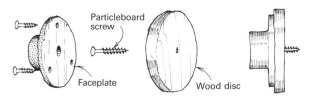

Particleboard screw

Faceplate

Wood disc

You can make all sorts of custom screw-drive centers for the lathe quickly and inexpensively by utilizing particleboard screws. These screws, which have only recently become commonly available, feature large, wide threads with tremendous holding power. If your local hardware store doesn't carry them, try the Woodworkers' Store, 21801 Industrial Blvd., Rogers, Minn. 55374-9514; (612) 428-2199.

To make a screw drive, mount a piece of wood to a small faceplate, turn a disc to the desired diameter and mark the center. Now remove the disc from the faceplate, drive the particleboard screw through the disc from the back and remount. The combination of the wide-thread screw and the wood-to-wood friction will give the drive good holding power.

—*Ken Picou, Austin, Tex.*

Taping square shoulders on turnings

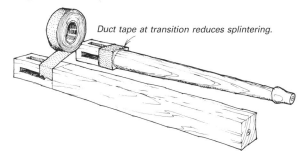

Duct tape at transition reduces splintering.

When your spindle pattern calls for a sharp transition from a square to round section, wrap the square section with duct tape so the tape edge defines the line of transition. The tape will reduce chip-out and provide an exact reference for making the shoulder cut. If chipping should occur, the tape holds the splinter for later regluing—no more searching through the pile for a lost splinter with less than a needle-in-a-haystack chance.

—*Robert M. Vaughan, Roanoke, Va.*

Using kitchen knives for lathe chatterwork

Years ago, I read an article in *Fine Woodworking* (#49) about chatterwork for decorating turnings. I was fascinated, and tried the technique of applying pressure against a thin spindle with a standard lathe tool until the workpiece would bend and chatter against the tool.

As you can imagine, this technique is risky, and I ruined a lot of expensive ebony before I discovered that it makes no difference whether the work chatters against the tool or the tool chatters against the work. So, I began making special chattering tools from thin stainless-steel kitchen knives. Although these

scrapers don't hold an edge long, they're flexible enough to produce beautiful chatterwork patterns. The patterns can be altered by approaching the work at different angles or by using different shape scrapers.

—*Ken Hopps, Tacoma, Wash.*

Wooden lathe chuck

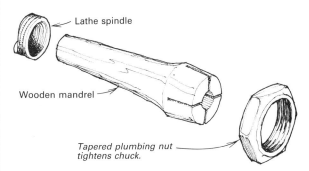

Lathe spindle

Wooden mandrel

Tapered plumbing nut tightens chuck.

I originally made this wooden lathe chuck to hold pieces of 1/4-in. dowel stock. But the design could be sized to fit any dowel or even to serve as a small collet chuck. To make the chuck, first turn a Morse taper on a piece of hard maple to fit your headstock spindle. Tap the future chuck into your headstock and turn a 1½-in.-long head on the end. The head should be tapered slightly and sized to fit a nut made by sawing an iron pipe bushing in two directly behind the hex flats. To complete the chuck, drill an accurate hole through the center of the head, using a bit in the tailstock center. Then make two opposing sawcuts along the hole to allow for compression. Insert the dowel and tighten the nut, and the dowel will be held firmly. Because pipe threads are tapered, be sure to install the nut large-end first. A little oil on the nut threads will help.

—*Walter O. Menning, La Salle, Ill.*

Turning from the left

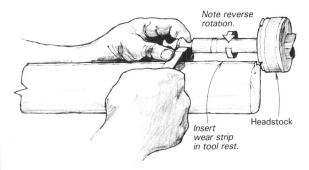

Note reverse rotation.

Headstock

Insert wear strip in tool rest.

This two-handed lathe technique works well for production turning small pieces. Note that the headstock is to the turner's right. The turner sits sideways, facing the headstock, his left arm over the tool rest with his elbow on the lathe bed. He holds the tool handle in his right hand, and with the left, guides the tool and steadies the spindle to prevent chatter.

A right-handed turner would have to turn his lathe 180° (headstock to the right) and reverse motor direction to use this technique. A left-handed turner would do this backward, with the headstock in its normal position and his right elbow resting on the lathe bed.

The full-length tool rest is made by inletting a wear strip in a long piece of wood.   —*Johannes Volmer, Erzgebirge, G.D.R.*

Driving with old engine valves

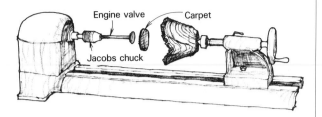

I've found that an old car valve works well as a live center for turning the base on a bowl that has its top edge left natural from the log. The valve stem is small enough in diameter to fit in a Jacobs chuck, and it's long enough to allow fairly deep bowls to be turned. The valves are usually free, because service stations that rebuild engines normally throw them out.

First, I fasten the valve in the chuck. Then with a small circle of indoor-outdoor carpeting inside the bowl for padding, I bring the ball-bearing tailstock up to the center of the bowl's bottom, which I've marked with a centerfinder. You need quite a bit of pressure for this technique to work properly, so leave plenty of wood on the bowl's bottom when you are roughing it out; you don't want the tailstock center to punch through the bottom. Things may get out of balance, so wear face protection and keep your lathe at low speed. Be sure to leave enough wood around the tailstock center for safety; this nub will be easy enough to clean up by hand after the bowl is removed from the lathe.

—Robyn Horn, Little Rock, Ark.

Wrench tenon cutters

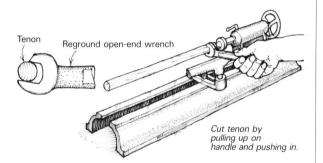

Cut tenon by pulling up on handle and pushing in.

Here's a quick production method for turning tenons: You can make a precision tenon cutter by modifying a high-quality open-end wrench of the same size as the desired tenon. First, carefully grind and sharpen the top jaw at an angle, as shown in the sketch, to provide a cutting edge. Add a handle to the tool if you like. Next, turn the stock to within 1/8 in. of the desired tenon diameter. Then, with the lower lip of the tenon cutter riding under the spindle, pull up on the handle and push in. The cutting action will stop when the tenon is sized—much like a go/no-go gauge. If the tool cuts tenons that are too small, file a bit off the lower lip. *—Cecil Gurganus, Todd, N.C.*

Turning splatter guard

There are several advantages to finishing a turned bowl or spindle on the lathe. But one big disadvantage is that the finish sprays all over the lathe and the wall behind. When I finally got tired of taping newspaper behind and on the lathe, I enlisted

the help of Dr. Bill Riddle, the metals instructor at our school, to build the finishing shield shown here.

We welded together pieces of 1/8-in. by 1 1/2-in. band iron for the shield's frame. The piece of band iron welded to the bottom

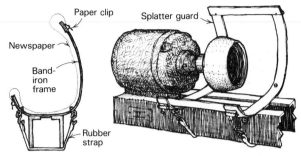

serves as an index to locate the shield between the lathe's ways. We riveted a No. 2 Boston paper clip to each corner to hold a sheet of newspaper spread inside the shield. Finally, we attached two rubber tie-down straps to the rear of the shield. These straps are pulled under the lathe bed and hooked into rings on the front of the saddle to cinch the shield down.

—Jerry Brownrigg, Alva, Okla.

Turning spheres

While chatting at a meeting of the Guild of Oregon Woodworkers, I learned that I turn spheres differently from most other turners. First, I turn a short cylinder with the grain running along the long axis. I turn the cylinder slightly longer than the diameter of the finished ball, then mark around the center of the cylinder with a pencil. Next, I square across one end and down each side to give me the location of two new turning centers. I chuck the cylinder in the lathe on these new centers and hang a light behind the turning. With the light turned on and the workpiece spinning in the lathe, the outline of a phantom sphere appears through the workpiece. I turn to this phantom line with a skew, being careful to avoid kickback, then sand and finish. If you don't have much experience with a skew, you may want to do most of the work with a gouge, then finish carefully with a skew.

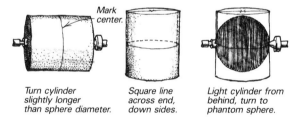

Turn cylinder slightly longer than sphere diameter.

Square line across end, down sides.

Light cylinder from behind, turn to phantom sphere.

This approach leaves tiny defects that can be filled and sanded where the centers have dented each side. Reduce these defects by using a small spur center and a ball-bearing center.

—Bill Fox, Salem, Ore.

Steady rest for baseball bats

For years, the eighth graders in my woodworking class have wanted to turn baseball bats. I've always put them off because the small diameter of the bat's handle invariably results in whipping and chattering, especially at the hands of an inexperienced turner. I licked the problem by making the steady rest shown here out of skateboard wheels, threaded rod and a couple of scraps of hardwood. The urethane plastic skateboard wheels can be bought for less than $15 a pair.

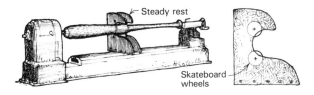

To use the steady rest, turn a cylinder, then tighten the steady rest in place with both wheels riding ahead of the handle. Turn the bat to shape, except the area right near the steady rest. My students spokeshave this area off, then sand and finish their bats.
—*Paul Damato, Morristown, N.J.*

Turning accurate tapers

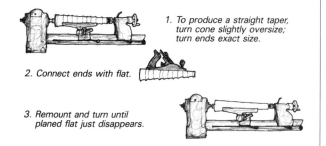

1. To produce a straight taper, turn cone slightly oversize; turn ends exact size.

2. Connect ends with flat.

3. Remount and turn until planed flat just disappears.

A recent request for a tinsmith's cone mandrel presented me with the problem of turning, freehand, an accurate taper. The technique I came up with is so simple and effective that I'd like to pass it on to other turners. First, rough out the stock slightly oversize and turn the ends to the final dimensions. Then, with the workpiece in a vise, plane a flat from the large to the small end until there is a straight taper all along. Now, re-center the turning in the lathe and turn the whole piece until the flat edge just barely disappears. The result will be an accurately tapered mandrel.
—*Tom Ryder, Sturbridge, Mass.*

Cheap faceplates
A bit of work will convert an inexpensive, common plumbing floor flange into a lathe faceplate for bowl turning. The biggest problem is that the threads on the floor flange are tapered. You'll have to use a tap in the appropriate size to open up the taper so the faceplate will screw on your lathe's spindle without binding. After you've opened up the threads, screw the flange on the spindle and check the fit of the hub against the shoulder of the headstock. File the high spots on the hub until it fits flat up against the shoulder. Now, with your lathe running at its slowest speed, scrape the face of the faceplate true with a carbide scraper or an old file ground into a chisel shape. To finish the faceplate, scrape or file the edge. If you use a file, be sure to keep it moving so you won't wear out one spot.
—*Robert Kelton, Saranac Lake, N.Y.*

Production turning with calipers

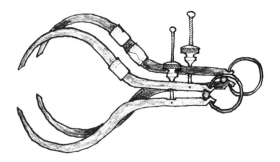

If you do much production-run spindle turning you have probably already discovered that it's much more efficient to use several calipers preset to various key diameters rather than to reset one caliper several times. But if you happen to pick up the wrong caliper at the wrong time it's easy to ruin the work. To reduce the chance of this error, I mark each caliper with bands of tape and set the calipers in the sequence used. The caliper for the first cut will have one band of tape, the second caliper two bands, and so on. This simple procedure has virtually eliminated mistakes.   —*Alan Dorr, Chico, Calif.*

Fixture for turning feet on bowls

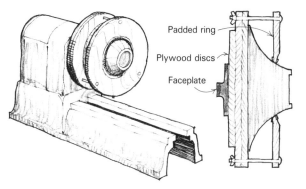

Padded ring

Plywood discs

Faceplate

This fixture is for those woodturners who appreciate a neatly turned foot on the bottom of a bowl or plate. Unlike other fixtures I've seen, it won't bust your knuckles, lose the workpiece at the critical time or require much fussing to center the work. The easy-to-build fixture is made from one 9-in. disc, two 12-in. discs and one 12-in. ring—all cut from good-quality plywood. Glue up the three discs into a solid base and mount them to a faceplate as shown in the sketch.

Now install three smooth-headed carriage bolts through the plywood ring into the base so that the work can be sandwiched between the ring and the base and tightened in place with wingnuts. It's a good idea to round and pad the inside of the ring so it won't mar the work. You may need several different rings, each with a different-size opening to handle various sizes of bowls and plates. To center work in the fixture, turn the work by hand as you tighten the wingnuts, adjusting the position as necessary—you can rest a pointed skew across the tool rest to use as a reference point.   —*Doug Napier, Mansfield, Ohio*

Turning hollow spheres

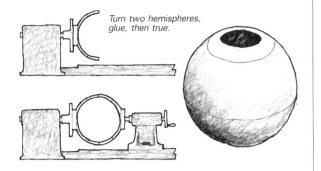

Turn two hemispheres, glue, then true.

For those of us who have neither the tools nor the skill to hollow out a solid sphere on the lathe, here is an alternative. First glue up two blanks using truncated wedge segments and solid wood caps (see "Segmented Turning," *FWW #54*). Turn two hemispheres, as shown in the sketch above, and glue them together after they have been hollowed to the desired wall thickness, leaving some extra thickness to allow for truing later. To glue the hemispheres together I leave one hemisphere attached to its faceplate in the lathe and use the tailstock to apply pressure while the glue sets. It is a good idea to leave the tailstock in place for extra support while you true the sphere to its final shape.

—*Al Brotzman, Madison, Ohio*

Adjustable go/no-go lathe gauges

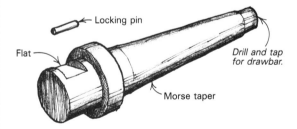

One full turn of screw changes gauge size by 1/32 in.

A set of these shopmade gauges will make measuring diameters easy and precise. Start with a length a 3/8-in.-thick, close-grained hardwood. Cut U-shaped recesses in each end slightly larger than the diameter to be measured and trim the piece so that the "arms" of the gauge end up about 3/8 in. square. Next, using a 9/64-in. bit, drill holes for the adjustment screws in each arm as illustrated. These holes should be 1/4 in. from the ends, square and centered.

Now install four 1-in.-long, size 8-32 machine screws in the holes. Turn the screws right into the holes—no tapping is necessary. In fact, the self-cut threads will make lock-nuts unnecessary. Round over and smooth the ends of the screws to increase accuracy. Now you're ready to set the gauge to the required dimension with an inside caliper or any convenient standard. One full turn of the screw makes a 1/32-in. change in the gauge dimension.

—*R. H. Taylor, Southport, Conn.*

Lathe sizing tool

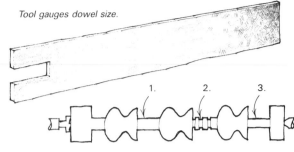

Tool gauges dowel size.

I turn many wooden knobs in multiples of four, and I needed a quick and accurate procedure to put 1/2-in. dowel stems on each knob. This easy-to-make sizing tool does the job beautifully. Start with a 12-in. length of 1/8-in.-thick steel. Square the end and cut a 1/2-in.-wide slot 1 in. or so deep into the end of the bar. File the slot to exact size. You may wish to bevel the tool's cutting edges above and below the slot, so they have a shape similar to the cutting edge of a parting tool. However, I found that the sizing tool will function perfectly well with sharp, square corners. To use the tool, first turn the knobs to shape with their stems slightly oversize (1). Plunge the sizing tool down on the stems at 1/4-in. intervals to produce bands of the true dowel size (2). Then bring the remainder of the dowel down to size with a parting tool or chisel (3).

—*J.C. Collier, Upper Hutt, New Zealand*

Self-locking pin chuck

This lathe chuck features an ingenious self-locking mechanism that allows quick and easy mounting and dismounting. It works equally well in both forward and reverse rotation. The chuck is ideal for projects with predrilled, centered holes, such as candlesticks, bud vases, wooden flutes and the like. You simply mount the hole over the end of the chuck to turn the profile.

To make the chuck, start with a length of mild steel bar. Turn a Morse taper on the tail of the chuck to fit your headstock. Then turn the head of the chuck to fit a predrilled hole in your turning blank—3/4 in. for example. Now file a flat spot on the head, as shown in the sketch. The depth of the flat should be just a bit greater than the diameter of the locking pin. The locking pin is nothing more than a piece of nail almost as long as the flat.

Locking pin
Flat
Drill and tap for drawbar.
Morse taper

To use the chuck, first drill a hole the same size and depth as the head of the chuck in your workpiece. With the locking pin centered in the flat, slip the workpiece on the chuck and rotate the work until the pin wedges and locks the workpiece in place. The chuck will lock in either direction—be sure you lock the work opposite the way your lathe will be turning. If you don't, tool pressure will unlock the chuck while you work.

—*John G. Martin, Cumberland, Maine*

Gaining length on a lathe

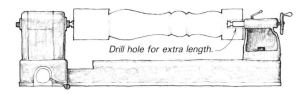

Drill hole for extra length.

When I wanted to turn a set of fancy 6x6 Victorian newel posts, I found my lathe's capacity was 2 in. short of the required length. Finally I discovered the method illustrated here, which gave me the extra 2 in. and paid a safety bonus as well. I drilled a hole the same diameter as the tailstock, 2 in. deep in the center of one end, dropped the tailstock into the hole and mounted the work on the lathe. After the workpiece is in place for turning, it's impossible for it to fly off the lathe.

—*Dan Miller, Elgin, Ill.*

Ferrule tool

In *FWW #45*, James Dupler described a lathe tool for turning beads. It reminded me of the similar tool I use for truing the ends of shopmade copper or brass ferrules.

Make the tool from a short length of ⅜-in.-sq. tool steel—the blade should project no more than 2 in. from the handle to reduce vibration. Grind the end to a diamond-shaped face.

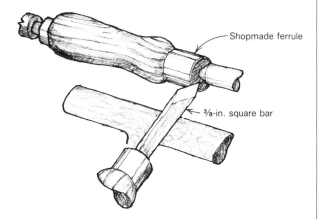

Shopmade ferrule

⅜-in. square bar

To use the tool, first turn a handle on the lathe. Remove the handle and drive a short section of copper tubing onto the end. Remount the handle, and with the toolrest close, bring the tool to the ferrule with the diamond face up. Roll the tool until the edge cuts, then proceed to level and round the end of the tubing.

—*P. W. Blandford, Stratford-on-Avon, England*

Compression rings for split turnings

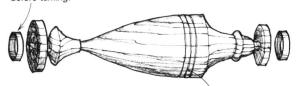

Drive rings on blank before turning.

After turning, remove rings and split turning apart at paper joint.

Half-round turned columns and finials make attractive decorative elements on clocks and chests. These are usually made by gluing up a laminated turning blank with paper between the pieces. After turning, the halves are separated by inserting a thin knife into the paper joint.

One drawback with this technique is that the lathe centers can wedge apart the weak paper joint when the blank is tightened on the lathe. To avoid the problem I use compression rings, driven in each end of the workpiece, to hold it together during turning. I make the rings from thin-wall tubing (conduit) by sharpening one end with a file, then I drive the rings about ¹⁄₁₆ in. into each end of the turning.

—*Norman Brooks, Greenville, Pa.*

Lathe-based sharpening wheel

Some time ago I decided to reshape my lathe skew chisels to Mike Darlow's specifications *(FWW #36)*. I devised this simple-to-make grinding wheel that uses the lathe itself to produce the 8-in.-dia. hollow grind Darlow recommends.

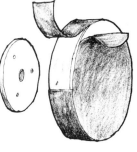

First, glue two 9-in.-dia., ¾-in.-thick plywood discs together and permanently screw them to a faceplate. Mount this on the lathe and turn the edge and face true.

Laminate a piece of ⅛-in. acrylic plastic to the face and smear a ⅛-in.-thick coat of epoxy around the full edge. When the epoxy is hard, turn the edge until it is true and flat, leaving as much epoxy as possible. True the face of the disc if needed. Using sanding-disc adhesive, glue coarse emery cloth to the face and edge of the disc to complete the grinding wheel.

To use the wheel, set the lathe at its slowest speed and rest the tool to be sharpened on the tool rest. Use the edge of the wheel for a hollow grind and the face of the wheel for a flat grind.

—*Dwight G. Gorrell, Centerville, Kan.*

Third hand for spindle copying

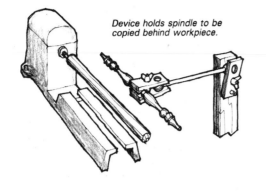

Device holds spindle to be copied behind workpiece.

This simple device, by holding a master spindle in full view directly behind the workpiece, eliminates much of the tedious measuring and template making that's usually required to duplicate a turned spindle. With the master copy registered near the work you can accurately judge lengths, critical layout cuts and even diameters and shapes by eye.

To make the device, turn a foot-long dowel "arm" with 1-in. balls on each end. Make up two pinch-blocks, as shown in the sketch, to lock the arm at any setting needed. The rear pinch-block may be attached to the lathe bed or fixed to a floor stand behind the lathe.

—*A. D. Goode, Sapphire, New South Wales, Australia*

Toolrest height stop

When you need to maintain one height setting for your lathe's toolrest, but have constantly to change its angle (as when faceplate turning) tighten a small hose-clamp on the toolrest's shank. This will prevent it from slipping down as you adjust it.

—*Brian P. Mitchell, Somerset, Colo.*

Faceplate centering device

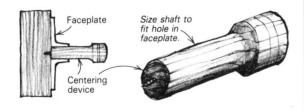

Faceplate

Size shaft to fit hole in faceplate.

Centering device

This simple little device will help locate a faceplate over the center of a workpiece. To use, first center-punch the workpiece. Then screw the centering device into the workpiece through the center hole of the faceplate to hold it in position while you drill the pilot holes for the fastening screws.

—*W. I. Newcomb, Arlington, Va.*

Flexible drum sander

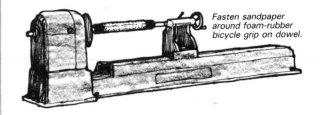

Fasten sandpaper around foam-rubber bicycle grip on dowel.

This inexpensive drum sander can be made by slipping a foam-rubber bicycle grip over a 5/8-in. dowel. The grip will stay nicely in place without adhesive. Notch and center-punch the dowel so it will run between centers on your lathe. Form sandpaper into a cylinder, scrape the abrasive off the bottom edge of the seam and hot-glue the sandpaper around the foam rubber. The sander has just enough give not to sand flat spots on curved surfaces but is firm enough to make smoothing fast and easy. I use a vacuum and homemade sawdust collector to pick up the dust. —*Gene Austin, Blue Bell, Pa.*

Patching turned spindles

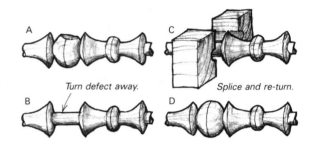

Turn defect away.   Splice and re-turn.

If you mess up a detail in an otherwise good turned spindle, you can cut a small block of similar wood and use it to patch the work, as shown in the sketch. The procedure can also be used to create unusual effects with contrasting woods.

—*John Sillick, Gasport, N.Y.*

Collet chuck for turning miniatures

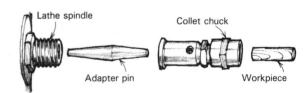

Lathe spindle   Collet chuck

Adapter pin   Workpiece

This collet chuck, adapted from a drill-press attachment, is inexpensive but effective for turning miniatures on the lathe. To make the chuck, purchase a drill-press attachment called a collet chuck with heavy collar (Sears part No. 9-24672). The holding collar that fastens the device to the drill press is not needed on the lathe. You'll also need an adapter pin with a male No. 33 Jacobs taper on one end and a Morse taper on the other end to fit your lathe spindle.

Fit the chuck to the lathe, then tighten the collet on a short length of 1/2-in. dowel, and you're ready to turn. A set of bushings that comes with the chuck will let you turn 1/4-in. and 3/8-in. dowels in addition to the nominal 1/2-in.

—*R. E. Hollenbach, Livermore, Calif.*

Wooden bearings for outboard lathe

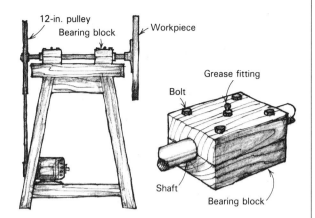

Several years ago, I needed an outboard faceplate lathe for turning large plates. While considering ways to home-build one, I remembered that as a young fellow I had helped my wife's dad as an oiler on a threshing machine. The contraption had a number of low-RPM shafts, which turned in hard-maple pillow blocks.

Adapting the idea to the project at hand, I purchased a 20-in.-long, 1-in.-thick shaft, fitted a 12-in. pulley (reclaimed from a clothes dryer) to one end, and threaded the other to accept standard faceplates. The shaft runs in two hard-maple pillow blocks, which are lubricated through grease fittings installed in the top.

The whole arrangement is bolted to a sturdy bench, and is run by a motor and belt from below.

—*Vic Johnson, Lincoln, Neb.*

Extending lathe capacity

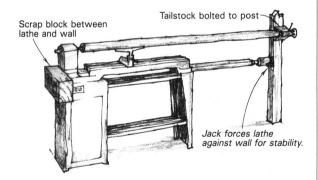

Here's how I extended the bed capacity of my lathe. First I bolted the tailstock to a support post in my shop. To make sure the tailstock was level with and in line with the headstock, I ran a chalkline and a line level between the two. To keep the lathe in position, I used a hydraulic cylinder to force the lathe's base against the wall—a spare Lally column would work nearly as well. A piece of scrapwood protects the lathe where it presses against the wall.

For slightly shorter stock, you could shim the lathe out farther from the wall, or dispense with the jack by bolting the lathe to the floor where you need it. I did half the turning and then flipped the workpiece to finish, so I could use my regular tool rest. If you have a freestanding tool rest, you can do the work all in one shot. —*D. Mayerson, Berkeley, Calif.*

Chucking bowl blanks

Many turners spend more time fiddling with faceplates and attaching the work to the lathe than actually turning. I'd rather spend my time turning, so I devised this quick procedure that takes me from a blank to a finished 10-in.-dia. bowl with 1/8-in.-thick walls in 30 minutes.

The key to the method is a 6-in-1 Universal Chuck, which has an expanding collet that locks into a dovetail recess in the workpiece. First I cut a recess in the top of the blank, using a router with a dovetail bit and the circular template shown in the sketch. The router rides around inside the shoulder on the template to produce a recess to fit the chuck.

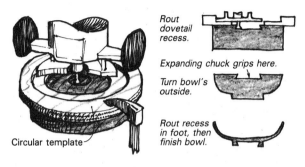

With the circular blank mounted on the chuck, I turn the bowl's outside profile. At this point you can turn a chuck recess in the bottom of the bowl if you choose, but I find it easier and faster to remove the bowl and cut the new recess with the router. It's important to center the bowl's foot in the template before cutting the recess, or the bowl will wobble on the lathe. If you turn the foot to fit the center hole in the template, this won't be a problem.

Now I return the bowl to the lathe and complete the inside. If desired, you can part off the bowl above the foot to eliminate all signs of the attachment method.

—*F.H. Crews, High Point, N.C.*

Foam faceplate for turning bowl feet

If your bowl design calls for a small foot (1½-in. dia.), here's a fast, easy procedure for chucking the bowl blank in the lathe.

First screw the bottom of the blank directly to a 3-in. faceplate and turn the inside to finished size. Turn the outside rim to size, but leave the bottom oversize so you won't hit the faceplate screws. Remove the blank from the faceplate and reverse it on

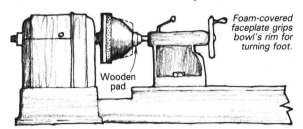

the lathe, holding it in place between a foam-covered faceplate and the tailstock. Now finish the foot to final size, cutting away all traces of the screw holes.

To make the faceplate, glue 1-in.-thick foam to a trued-up 4-in. or 5-in. maple disc screwed to a 3-in. faceplate. I use a ball-bearing tailstock center, fitted with a 7/8-in. flat wooden pad, to press the bowl into the foam disc. —*Max M. Kline, Saluda, N.C.*

Bowlturning chuck

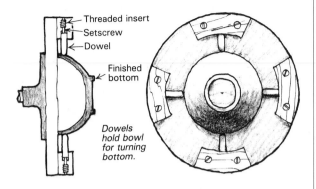

Threaded insert
Setscrew
Dowel

Finished bottom

Dowels hold bowl for turning bottom.

I make bowls by turning the top and inside first, then reversing the blank and turning the bottom. This lets me use a standard faceplate for the heavy roughing-out and hollowing operations. For the second step, I switch to a special chuck to finish the bottom. The shopmade chuck described here does a good job—four dowels grip the bowl's rim and provide adjustment for centering. To make the chuck, mount a ¾-in- thick, 12-in-dia. disc to a faceplate, true it and mark the center. Remove the disc and screw four 1-in.-thick, 2-in.-wide segments to the rim 90° apart. Return the disc to the lathe and true the segments into semicircular arcs 1½ in. wide. Remove the disc again and mark the centerline of each segment radially for installing a threaded insert. Counterbore each segment from the inside (remove if necessary) to accept a ⅜-in. or ½-in. dowel pin. Screw hex-head setscrews in the threaded inserts to tighten the dowels against the bowl rim.

To use the chuck, first mount the bowl on a faceplate, and turn and sand the top and inside. While the bowl is still on the faceplate, mark the center of the bottom with a pointed steel rod through the back of the faceplate.

Remove the bowl from the faceplate and mount it in the special chuck. To center the bowl, bring up the tailstock and use the point on the dead-center as a reference. Tighten the work in the chuck by screwing in the setscrews in the rim, then retract the tailstock. With longer dowel pins, the chuck will hold work as small as 4 in. Of course, the chuck could be scaled down for smaller work.

For safety's sake, limit your work to the very bottom of the bowl—keep your fingers away from the exposed dowels.

—F.K. Anan, Tokyo, Japan

Homebuilt outboard lathe

Turning circular tabletops on my regular lathe was less than satisfactory. The outboard spindle was just not designed for large, unbalanced, rough work. When a friend offered me a rear wheel and axle bearing from a front-wheel-drive car (G.M. No. 1-7466906), my ideas for a special homebuilt outboard lathe came together. I figured that if the hub could handle a car wheel, it would be ideal for turning a tabletop.

I bolted the wheel assembly's brake flange to a 12-in.-long section of ¼-in- thick, 3x3 angle iron as shown, and lag-screwed this to a rigid yellow-pine bed about 5 ft. long.

The lathe faceplate is a 1-in.-thick, 11-in.-dia. oak disc. I bolted the faceplate directly to the hub with a 9-in. pulley sandwiched between. The headstock/pulley assembly is permanent, and after installation the faceplate should be trued round and faced flat.

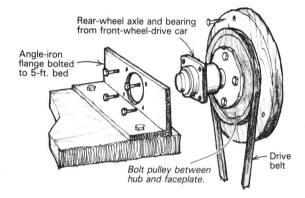

Rear-wheel axle and bearing from front-wheel-drive car

Angle-iron flange bolted to 5-ft. bed

Drive belt

Bolt pulley between hub and faceplate.

To power the lathe, I mounted a ½-HP, 1725-RPM motor with a 2-in. drive pulley.

The easiest way to fasten the work to the lathe headstock is to drive screws through the rim from the back side. Of course, more elaborate faceplate-fastening techniques can be designed for special projects if needed.

Even on the first project, the lathe exceeded my expectations with its quiet, vibration-free performance.

—Lawrence Wachenheim, Quincy, Ill.

Router Jigs and Fixtures

Chapter 12

Plate joinery on a budget

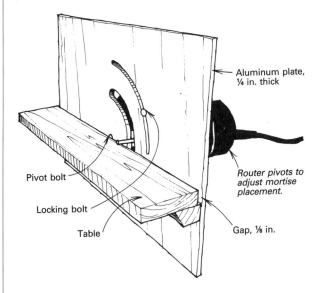

Three-wing slotting cutter

Jig guides cutter in arc to fit radius of biscuits.

When my hankering to take advantage of quick biscuit-joint systems ran up against the high cost of the required machinery, I looked for a cheaper approach. My solution was to use a wing slotting cutter in a router to make the kerfs for the standard biscuits. An Amana three-wing slotting cutter (available from W.S. Jenks & Sons, 1933 Montana Ave., N.E., Washington, D.C. 20002 for about $15) can cut a 5/32-in.-wide, 1/2-in.-deep slot the same as any biscuit joiner.

The only problem is that the three-wing cutter's radius is just under 1 in., while the radius of the biscuit joiner's sawblade is 2 in., so the profile of the routed slot will not mate perfectly with the semicircular edge of the biscuit. Although this mismatch will not affect the assembly or strength of the joints, it can be eliminated by constructing a simple jig, as shown in the sketch. The jig guides the cutter through a 2-in. arc and also sets the depth of cut.   —*Richard Fryklund, Arlington, Va.*

Pivoting router mortising fixture

Because I needed to cut more than 400 mortises in a short period of time, I built this pivoting router fixture. With it, I can cut two mortises in about one minute, including the layout time, so the four hours I spent building the jig were quickly repaid.

The router is attached to an aluminum plate with a single bolt so it will pivot to adjust for the position of the mortise in the stock. The plate has two concentric slots centered on the pivot bolt: one for the mortising bit and one for a locking bolt and wing nut. A cleat to support the table is screwed to the plate and a hardwood table is glued and screwed to the cleat with a 1/8-in. gap left between the table and plate for chip and dust clearance. My aluminum plate is 1/4x12x20. I recommend 6061 aluminum with a hardness of at least T3. You can mill the curved

slots in the plate by building a special pivoting fixture and using a milling cutter in the drill press. Or, if you're patient and careful, you can rout the slots with a router and double-flute carbide bits with a trammel or circle-cutting fixture. Take several light cuts. After the plate is completed, install the fence and attach your router.

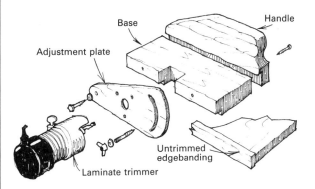

Aluminum plate, 1/4 in. thick

Pivot bolt

Locking bolt

Table

Router pivots to adjust mortise placement.

Gap, 1/8 in.

Bolt or clamp the jig to the edge of a stout table or workbench. Adjust the router for mortise placement and depth of cut. Then start the router and push the stock from left to right past the bit. Plunge the stock onto the bit for stopped mortises. Use stop blocks for repetitive cuts or draw layout lines on your stock to show you where to start and stop your mortise in relation to the bit's slot. Don't try to mortise pieces that are too narrow or are shorter than about 12 in. Use the same caution you would with any router-table operation.   —*James E. Gier, Pine, Ariz.*

Trimming edgebanding

Base

Handle

Adjustment plate

Untrimmed edgebanding

Laminate trimmer

Trimming solid-wood edgebanding on plywood with a plane or belt sander can be a trying task, so I designed this simple trimmer fixture, which holds a horizontally mounted router. An adjustment mechanism allows the router's bit to be adjusted up or down so it cuts the banding flush with the plywood. Instead of a full-size router, I use a Porter-Cable laminate trimmer, which provides plenty of power for trimming the 1/4-in.- to 3/8-in.-thick banding I use. I've found that an Onsrud 1/4-in., two-flute spiral bit gives a smooth, splinter-free cut that's ready for finish-sanding.

—*Warren W. Bender Jr., Medford, N.Y.*

Panel-raising fixture

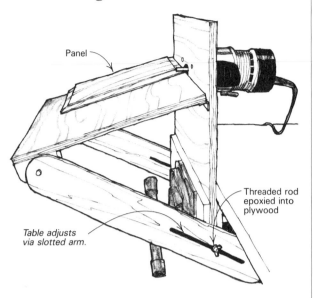

Panel

Threaded rod epoxied into plywood

Table adjusts via slotted arm.

This router fixture is great for quickly and safely beveling panels. It's made from ¾-in plywood and some scraps. Its table is hinged and adjusts via a slotted support arm to vary the bevel angle.

I use a ½-in.-dia. carbide bit and make three passes. The first two passes remove the bulk of the stock. The final pass removes only about ⅟₃₂ in. of material and leaves a clean face without ripples. —*Gerald Robertson, Angus, Ont.*

Single-setup routed drawer joint

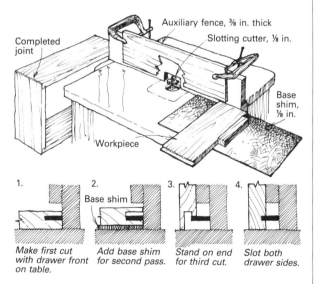

Auxiliary fence, ⅜ in. thick

Slotting cutter, ⅛ in.

Completed joint

Base shim, ⅛ in.

Workpiece

1. Make first cut with drawer front on table.
2. Add base shim for second pass. Base shim
3. Stand on end for third cut.
4. Slot both drawer sides.

Claude Graham's article on production drawermaking (*FWW* #72, pp. 82-85) prompted me to send this single-setup solution for routing tongue-and-rabbet drawer joints. The setup uses a standard ⅛-in. slotting cutter chucked in a router that's mounted under a router table. You'll also need a piece of ⅛-in.-thick Masonite for a base shim and a length of ⅜-in.-thick material for an auxiliary fence. Adjust the height of the cutter so the slot starts ½ in. above the router table, and adjust the fence so the slot's depth is a shade over ½ in. (this is usually the full depth of the cutter). Clamp the auxiliary fence to the main fence, ¾ in. above the table so a ½-in. drawer side and the base shim can slide underneath.

The sequence of cuts is shown in the sketch. Three passes will produce the tongue and rabbet on the drawer fronts and backs. One pass will produce the groove on the drawer sides. Although it's not shown, you can also use the same setup to groove the drawer sides to receive the bottom. Screw the auxiliary fence to the tool's fence so you can remove the clamps for clearance and make two passes, one against the table and another using the base shim. The result will be a ¼-in. groove ⅛ in. from the bottom of the drawer side.

—*Brad Schwartz, Deer Isle, Maine*

Making fixed-louver shutters

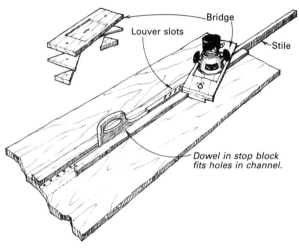

Bridge

Louver slots

Stile

Dowel in stop block fits holes in channel.

Exterior wood really takes a beating in the subtropical climate of Key West, Fla., where I work. So when a client hired our company to replace his deteriorated pine shutters with longer-lasting redwood, we developed this quick, easy and inexpensive method for building the shutters right on the job site.

A tedious but critical part of shutter construction is routing slots in the stiles for louvers. For this operation, we devised a jig, shown in the sketch above, that holds the stile in a channel and allows it to be moved in steps as each louver slot is routed. Accurate spacing is ensured by using a stop block and a series of holes 1⅜ in. apart in the channel. A dowel in the bottom of the stop block fits the holes. The jig is also fitted with a bridge that holds the router above and at 18° to the stile. The bridge may be unfastened and repositioned to cut mirror-image slots in the mating stile. A recess in the top of the bridge allows the router to travel back and forth the precise distance needed to cut each slot.

To rout louver slots in a stile, we first mark off 6 in. at the top and bottom to allow enough room for the rails and waste. We also mark off the center of the stile where we skip two slots to leave room for the middle rail. Then, with the stile located in the jig's channel, we begin routing slots, tipping back the router to start the cut (a plunge router would be great for this job). After repositioning the bridge, we rout slots in the opposite stile. To complete the shutters, we mortise the stiles, cut tenons on the rails and glue up with epoxy and pipe clamps.

—*Barb M. Kamm, Key West, Fla.*

Sliding dovetail jig

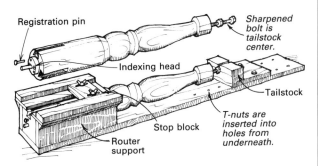

I use this jig for routing the sliding dovetail housings for the legs of small pedestal tables and stands: It's fast to set up and very accurate. I made the jig's index head from a ½-in.-thick aluminum plate bandsawn into a circle, but you can make it as easily out of a thick piece of hard maple. I tapped the center of the index head to receive a short length of ½-in.-dia. aluminum rod with one end protruding slightly and pointed to act as a center. Three indexing holes are bored through the head to correspond with the dovetail housings to be cut in the pedestal. A registration pin pushed through the face of the router support seats in one of these holes, positioning the pedestal for routing.

I use a ¾-in. dovetail cutter in my router to cut the housings and install an adjustable stop block on the router support to keep the housings the same length.

—*Eric Schramm, Los Gatos, Calif.*

Carcase dadoing jig

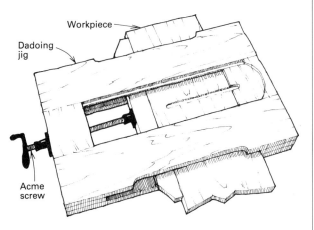

The homemade dadoing jig illustrated in Christian Becksvoort's article "Building a Chest of Drawers" (*FWW* #68) is similar to the one I made several years ago. Mine differs in one respect that I think improves the jig. I purchased an Acme screw, which is often used in constructing book, cheese and juice presses and is available through many tool catalogs. I installed the screw as shown to make the positioning and clamping of the jig quick and accurate—certainly better than using C-clamps.

—*Charles Leik, Great Falls, Va.*

Milling radiused corners on tabletops

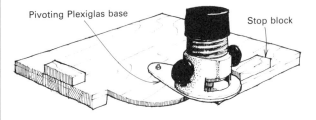

Faced with the prospect of milling 80 identical radiused corners on a run of restaurant tables, I came up with the "Corner King" jig shown in the sketch. It's built from a square of ¼-in. plywood, with fences attached to the bottom on two sides. A pivoting Plexiglas base was designed to allow a ½-in. router to swing through the proper radius (4 in. in this case). Adding stop blocks to the top limited the travel of the bit to 90°.

A nice feature of the jig is that the first pass with the router cuts the jig's base into a perfectly radiused pattern. In practice, I set the jig on a corner, traced the radius pattern directly off the base, removed the jig and trimmed the bulk of the waste with a jigsaw. Then I screwed the jig to the tabletop and used the router to finish the corner.

—*Al Dorsa, St. Croix, Virgin Islands*

Routing fingernail edges

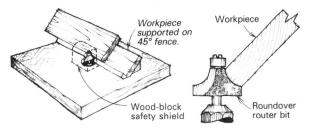

Your roundover router bits can do double duty cutting fingernail-shaped edges if you simply change the angle at which the work moves into the bit. I use the bits in a router table with an auxiliary fence that presents the stock to the bit at a 45° angle as illustrated. A ½-in. bit will mold the shape on ⅝-in.-thick stock, and a ¾-in. bit will handle 1-in.-thick stock. Notice that the lip on the fence acts as a track for the work and must have a gap in it so the bit can contact the work. For occasional use, this method beats buying the expensive specialty bit.

—*Jeffrey P. Gyving, Point Arena, Calif.*

Makeshift plunge routing

You can adapt a standard router with a screw-lowering mechanism to allow it to make plunge cuts. Fasten a hose clamp around the waist of the motor housing to stop the plunge cut at the desired depth. Loosen the router's tightening cam halfway and spray inside the base with Teflon lubricant.

—*James Gentry, Madison, Wis.*

Routing dado joints

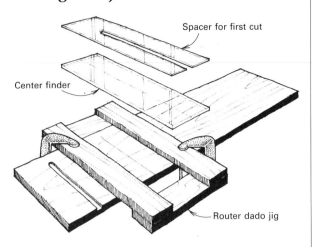

Spacer for first cut

Center finder

Router dado jig

In the high-school furniture-making class I teach, we use a router and a parallel guide like the one shown in the sketch to cut most of our dado joints. Even though the guide alone helps reduce errors, we use two simple plastic fixtures to increase accuracy and reduce the number of mistakes made by new woodworkers.

The first fixture is a clear plastic position finder, which we use to locate the guide quickly and accurately on the workpiece. To make one, cut the plastic the same width as your router base and as long as the guide. Then, scribe a centerline the length of the finder. To use it, first lay out the centerlines of the dados on your workpiece. Place the position finder in the guide, move the guide so the scribe line is positioned over the layout line and then clamp the guide in place.

The majority of our dado cuts are ¾ in. wide and ⅜ in. deep, which is too heavy a cut to make in one pass. Rather than reset the routing depth over and over for each cut, we use the second fixture, a ³⁄₁₆-in.-thick piece of plastic, as a spacer for the first cut. Like the finder above, cut the spacer the same width as the router base and the same length as the guide. Cut a 1-in. slot down the middle of the spacer to within a couple of inches of one end. To use, put the spacer between the guide fences, set the router for the full ⅜-in. depth of cut and make the first pass. Remove the spacer and make a second pass to the final depth.

—*J.K. Blasius, Bowling Green, Ohio*

Jig for sliding dovetail housings

I use a simple but effective jig to cut housings for sliding dovetails in drawer construction. The jig consists of an L-shaped shelf, a fence to guide the router and a spacer board screwed to the fence from the bottom. The jig is clamped to the front of the workbench from underneath with pipe clamps and is carefully adjusted so the height of the shelf matches the thickness of the drawer stock.

The jig is designed so that housings are cut ½ in. from the end of the workpiece. If necessary, adjust the size of the spacer to locate the housing farther from the edge. The grooves for the drawer bottom are cut in the drawer front and sides before the jig is used.

To use the jig, butt two sides up to the stop as shown, with the grooves at the far side of the drawer stock. Move the router in from the front of the jig, and stop the cut at the groove. To cut the housings in the drawer front, place the front so it faces in the opposite direction, with the bottom groove in front. Rout through the groove, stopping the cut for the housing at the desired distance from the top edge (usually ½ in. or so). This way, the sliding dovetail is not exposed at the top edge of the drawer's front.

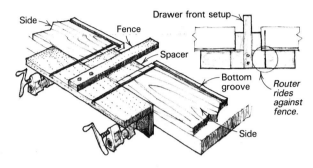

Side

Fence

Drawer front setup

Spacer

Bottom groove

Router rides against fence.

Side

To rout the male dovetails, I use a tall fence on my router table with the router attached to the back of the fence and the bit running parallel to the table. I recommend cutting one side of the dovetail on all the pieces, then resetting the fence and cutting the other side with the same face against the table as before. The principle of always working relative to one face will ensure that all dovetails will be the same size.

—*Barrie Graham, Arundel, Que.*

Making long dowels

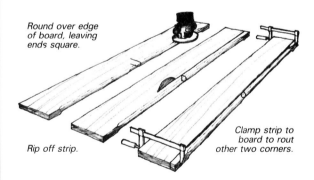

Round over edge of board, leaving ends square.

Rip off strip.

Clamp strip to board to rout other two corners.

I needed a ¾-in.-dia. oak dowel more than 6 ft. long to use as a curtain rod. Unable to find a source of supply, I came up with a method to make the dowel in my shop without a lot of effort. First, I rounded the edges of a 7-ft.-long, ¾-in.-thick oak board with a ⅜-in. corner-round bit in my router. I didn't round over the first and last 6 in. of the board because I knew I'd need the square ends for a clamping surface later. Moving the board to the tablesaw, I set my fence at ¾ in. and ripped off the rounded edge. I then flipped the stick over, clamped it back to the board with three quick-action clamps (for stability) and rounded the top and bottom edges again to produce a dowel. Of course, I had to reposition the middle clamp to finish the rounding. When I trimmed off the two square ends, I had my long dowel, ready to scrape, sand and finish.

—*Phil Lisik, Hemlock, Mich.*

Making tenons on chair rungs

Here's how to use your router table to produce tenons on the end of chair rungs, quickly and accurately. First, chuck a rabbeting bit into the router and raise it until the bit's bottom is even with the top of the router table. Locate a V-block near the bit to produce the diameter desired, and clamp the V-block in place using the router table's fence. Then, holding the rung firmly with one hand, lower it into the rotating bit. Rotate the rung counterclockwise with the other hand. The result will be a clean and uniform reduction of the dowel diameter. To reduce splintering, take several small bites of 1/8 in. or less. —*David J. Langley, Corvallis, Ore.*

Routing tablesaw inserts

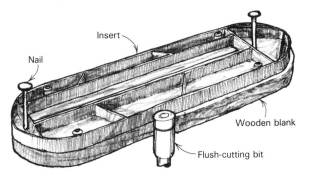

Drilling two nail-sized holes in your tablesaw insert lets you tack the insert to a rough-cut blank and pattern-rout a replacement wooden insert that's exactly the size of the original. A flush-trim bit with a ball-bearing pilot works best for the routing. Before you start, thickness-plane the stock for the blanks to the exact depth of your insert hole. I make the inserts up by the dozen and put in a new one at each blade change.

—*Jeffrey P. Gyving, Point Arena, Calif.*

Crank adjustment for router table

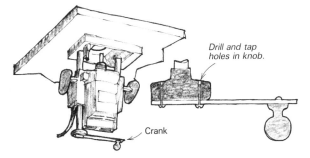

The Makita plunge router I installed in my router table works beautifully, but adjusting the depth of the bit with the adjustment knob was awkward and tedious. I solved this problem by removing the pre-load springs from the router support tubes. Then I fashioned a simple crank handle that screws to the existing knob. Now I can adjust the router depth quickly.

—*Robert T. Combs, Carpinteria, Calif.*

Self-made mortising template

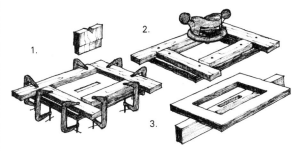

This procedure for making a router template is quite accurate because in the early stages the template uses itself for the setup and quality control. First lay out the mortise dimensions on the template stock. Now, with the router and bit you intend to use for the actual mortising, line up the cutting circle of the bit with one wall of the mortise. Clamp a strip parallel to the mortise side so it butts against the router base, thus defining that mortise wall. Repeat the process on the other three sides.

Now, as a test, rout a shallow mortise in the template stock. If the tenon does not fit, move and reclamp the guide boards. If the mortise is slightly oversized, you can add shims. Then cut another test mortise, a little deeper, and repeat until you have the fit you want. Next screw the guide strips in position, countersink the screw heads and remove the clamps.

To finish the template, cut out the center of the blank and turn the edges flush with a router and a flush-cutting bit as shown in step 2, then remove the guide strips.

—*Patrick Warner, Escondido, Calif.*

Routing tambour grooves

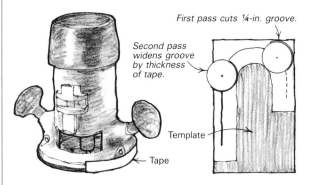

I recently built a set of display cases that had tamboured doors with 1/4-in.-thick edges. I wanted the grooves in which they ran to be 1/32 in. wider so the tambour wouldn't bind. To accomplish this I applied iron-on veneer edging tape around half the radius of my router base. To cut the groove I ran the router base along a template using a 1/4-in. straight bit. On the first pass I kept the router's original base against the template. On the second pass I rotated the router so the taped portion of the base bore against the pattern, thereby adding about 1/32 in. to the groove width.

—*Andrew Dey, Wallingford, Conn.*

Shaping beams with a router

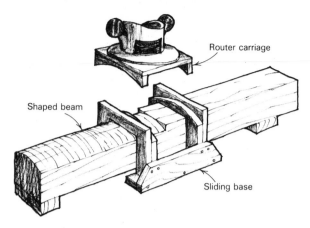

Router carriage

Shaped beam

Sliding base

Four years ago my eight-year-old daughter, an aspiring gymnast, pleaded for a balance beam of her own to practice on. Her request required me to find a way to shape the sides of a 16-ft. beam into uniform arcs, so that the finished beam would be as near regulation size and shape as possible. My solution was a sliding jig that guided a router with a 1-in. bit. The jig consists of two parts: the sliding base and the router carriage. Curved rails on these parts guide the router in the proper arc. In laying out the jig you must increase the radius of the curved rails by the amount that the bit protrudes from the router base so that the end of the bit follows the desired finished radius. I recommend you lay out the plan of the jig full-size to verify the correct juxtaposition of beam, cutter and jig.

To use the device, start at one end of the beam and arc the router to and fro as you slide the jig along. The router will let you know how much of a bite to take. The process is slow but accurate. To finish up the very ends of the beam, where the bit can't reach, you can rig up some additional bearing surface or simply use a chisel and plane.

—*Burt Babkes, Eugene, Ore.*

Centering routed mortises

John Birchard's door-making article *(FWW #49)* prompted me to send this mortise-centering idea I've used for quite awhile. Like Birchard, I use a plunge router to cut mortises in the stiles of frames. However, instead of using a fence to center the mortise, I attach two small ball bearings under the router base. On the Hitachi router I use (and most other routers), the subbase is attached with four screws. I remove two diagonally opposite screws and replace them with the bearing shown in

the sketch. The bearing rides on a shopmade press-fit insert that is slightly longer than the bearing's thickness, and that has an inner diameter to fit the bolt. The flange at the end of the insert can be machined as part of it, or it can be a separate washer. It need only be thick enough to prevent the bearing from rubbing on the router base.

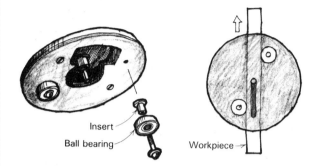

Insert

Ball bearing

Workpiece

In cutting the mortise, the bearings ride against the sides of the stile, automatically centering the mortise in the work. The size of the bearings is unimportant, as long as both are the same. Note that when cutting mortises near the end of the stile you must have an excess length, a "horn," for the bearing to ride on. Leaving a bit of excess to be trimmed off later is good practice anyway.

—*David Ring, Yodfat, Israel*

Routing V-grooves in tongue-and-groove

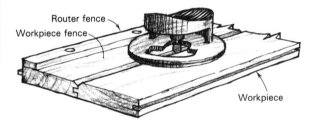

Router fence

Workpiece fence

Workpiece

To produce identical chamfers on matching edges of tongue-and-groove stock, I use an extra piece of stock with a nailed-on router fence, as shown in the sketch. Both the tongued and the grooved edges can be pushed flush to the jig, ensuring a balanced V-groove in the finished work and eliminating the extra setup that would be required with a shaper or a tablesaw. You could adapt the idea to a router table just as easily.

—*W. A. Ward, Underhill, Vt.*

Other Power Tools, Jigs and Fixtures

Chapter 13

Thickness-planing short pieces

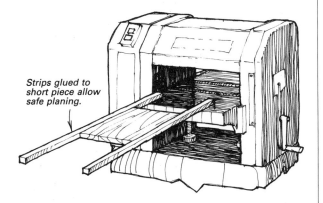

Strips glued to short piece allow safe planing.

Because feeding short boards through the planer may result in snipes and even kickbacks, the operation should be avoided. But when the job is necessary, here's a way to make it less risky. Glue two scrap outriggers to the edges of the piece to be planed, as shown in the sketch. These scraps, because they extend several inches beyond the ends, will stabilize the short board as it enters and leaves the planer, thus reducing the chance of sniping. When the desired thickness is reached, saw off the scrap outriggers and run the board's edges over the jointer to clean them up.   —*Bill Clark, Bakersfield, Calif.*

Raised panels on the jointer

You can produce panels with beautiful, long tapered bevels on your jointer, provided it has rabbeting capability. Build an outboard table from ¾-in. plywood about 12 in. wide and as long as your jointer. Mount the table to your jointer or jointer stand so that the outside edge may be raised or lowered to produce the desired angle. The inboard edge of the table should be about ¼ in. from the jointer's cutterhead.

Clamp a strip of wood to the jointer's fence flush with the jointer's bed to prevent the thin edge of the panel from sliding under. Three interrelated factors determine the shape of the bevel. The depth of the shoulder on the panel is determined by the height at which you set the infeed table. The distance from the fence to the end of the cutterhead determines the width of

the bevel. The thickness of the edge of the panel is determined by the angle of the outboard table, the width of the bevel and the depth of the shoulder.

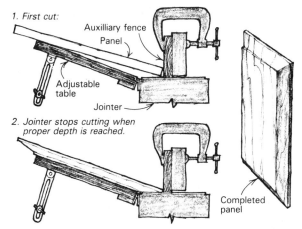

1. First cut:
Auxilliary fence
Panel
Adjustable table
Jointer

2. Jointer stops cutting when proper depth is reached.

Completed panel

When the setup is right, turn on the jointer and slide the panel over the cutterhead. Since the shoulder will ride on the jointer's infeed table when the proper depth is reached, you simply continue making passes until the jointer's knives stop cutting. Cut the two end-grain sides of the panel first so that tearout will be removed when the other two sides are cut.
                                                    —*Norris S. White, Sellersville, Pa.*

Wooden spring for outfeed support

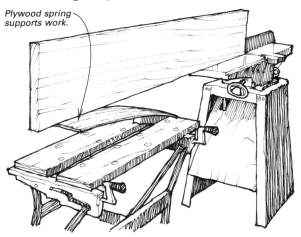

Plywood spring supports work.

I tried using a roller work-support while jointing the edges of long, heavy boards, but no matter how carefully I adjusted the height, invariably the work would be marked when the end of the board bumped the roller. What I really needed was a more flexible support that simply helped hold up the front of the board. The arrangement I came up with consists of a 2-ft.-long, ¼-in. plywood spring screwed to a 6-in.-long 2x4 clamping spline. I clamp the spring in a portable vise and adjust it so it balances the weight of the work and requires only a slight downward pressure to keep the work flat on the outfeed table.
                                                    —*Jack Jerome, Nokomis, Fla.*

Jointer thicknessing—another design

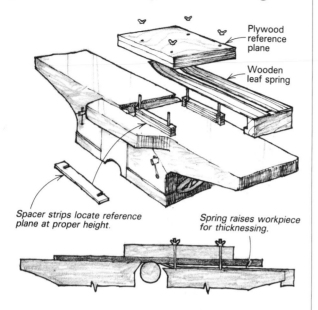

Plywood reference plane

Wooden leaf spring

Spacer strips locate reference plane at proper height.

Spring raises workpiece for thicknessing.

The design presented here converts a jointer to a true thickness planer, and it takes just a couple of minutes to make the conversion. On a regular thickness planer, the work is fed between the cutterhead and a flat bed. My method's principle, if you imagine a regular planer turned upside down, is the same: the workpiece is raised by wooden springs and pressed against a rigid overhead reference plane above the cutterhead.

My reference plane is laminated from two pieces of ¾-in., 9-ply birch plywood. It is held above the cutterhead by an assembly of aluminum spacers, threaded rods tapped into the feed table, and wing nuts. Another necessary component is a flat leaf spring, which presses the workpiece against the reference plane until it begins to ride the outfeed table. I made the spring from two 2½-in.-wide, ⅛-in.-thick strips of oak by steambending the last 4 in. of the strips. A simple cleat that hooks onto the end of the jointer table holds the spring in place.

The dimensions of the reference plane are not critical—an 8-in. square is about right for a 6-in. jointer. Wax the underside of the plane to reduce friction. I made the spacer strips by epoxying layers of ⅛-in.-thick aluminum strips together to form pairs of ¾-in., ½-in., ¼-in. and ⅛-in. spacers. The spacers are notched so they can slip onto the threaded rods with the reference plane in place.

To use, joint one face of all the pieces you are going to plane. Select spacers equal to the thickest piece plus another ¼ in. for the spring. Bolt the assembly in place, position the spring, and back the infeed table down until the thickest piece starts to cut. Plane all the pieces at this setting before lowering the infeed table for the next cut. Continue until the final thickness is reached.   —*J. E. Keister, Cincinnati, Ohio*

Thickness-planing on the jointer

Tage Frid, in *FWW* #19, p. 94, describes how to thickness boards on the jointer. Frid's jig is a precision wooden affair that requires removing the jointer's fence to work. Here's a simpler way. From a signmaker obtain two 1-in. strips of flexible magnetic sign backing and glue each to a hardwood strip to

produce two ½-in. thick sticks as long as the infeed table. Glue a hardwood block on the end of each strip to keep it from creeping into the cutterhead.

Before using the setup, first joint one face and both edges of the board to be thicknessed. Rabbet the edges, as shown on the workpiece in the sketch.

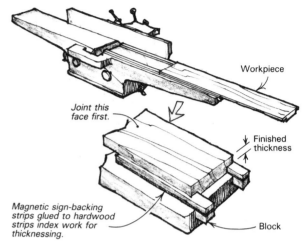

Workpiece

Joint this face first.

Finished thickness

Magnetic sign-backing strips glued to hardwood strips index work for thicknessing.

Block

Now snap the two strips in place on the infeed table so the rabbets ride the strips like rails. Run the workpiece down the rails, across the cutterhead and onto the outfeed table. In this manner, it is the uniform rabbet that indexes the work; the irregular face doesn't touch the infeed table at all. Start with a light cut, then gradually lower the infeed table with each pass until the rabbets are only 1/16 in. deep. On the last pass, just skim off the wood down to the rabbets to produce the final thickness.

The magnetic strips can be easily adjusted to different-width boards, and there's no need to remove the jointer's fence to use them. When the job is done, it takes all of three seconds to convert your thickness planer back to a jointer.

—*Robert Edmondson, Bowmanville, Ont.*

Hemispherical sander

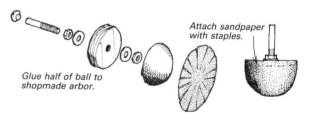

Attach sandpaper with staples.

Glue half of ball to shopmade arbor.

This little device, which works especially well for sanding concave interior surfaces, can be easily produced in the workshop. Start with a sponge-rubber ball, the kind available at toy stores in various diameters from 1 in. to 3 in., and carefully cut the ball in half. Now saw the head off a ⅜-in. carriage bolt. Using jam nuts, screw the bolt to a plywood disc the same diameter as the sponge-rubber hemisphere. Hollow the hemisphere to accept the jam nut and attach it to the plywood disc with hot-melt glue or silicone adhesive. Prepare an abrasive disc to conform to the hemisphere by cutting several radial slots at equal distances, as shown. Staple the abrasive paper to the wooden disc or hold the paper in place with a hose clamp.

—*Donald F. Kinnaman, Phoenix, Ariz.*

Spindle tapering jig

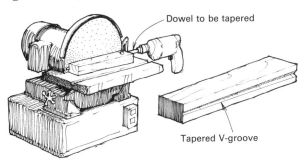

Dowel to be tapered

Tapered V-groove

Recently I needed a quantity of tapered wood rods to make drop spindles for spinning wool. The spindles' diameter had to taper from ½ in. to ¼ in. evenly along their 12-in. length. Rather than attempt to turn these tiny spindles on the lathe, I designed the jig shown above, which worked perfectly.

First I selected a 12-in.-long chunk of 1½-in.-thick hardwood for the sanding guide and sawed a ½-in.-deep V-groove into one edge. After cutting the groove, I ripped the guide at a slight taper so the groove was only ¼ in. deep at one end. To use the guide, I clamped it to the worktable of a disc sander, with the grooved edge almost flush against the disc. Then I chucked a length of dowel in an electric drill, and with the drill at slow speed, guided the dowel into the V-groove. After a little practice and some trial-and-error setting of the guide, I was able to make perfect tapers every time.

—*Bert. G. Whitchurch, Hemet, Calif.*

Coarse and fine sanding on the same disc

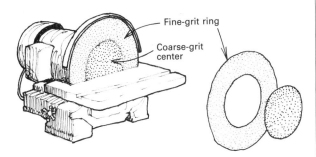

Fine-grit ring

Coarse-grit center

On some of the work I do on my 12-in. stationary disc sander, I often need to switch between 60-grit sandpaper for fast stock removal and 120-grit for finish-sanding. Changing the paper is a chore, and sometimes the paper is ruined in the process. For efficiency, I decided to try this two-grit arrangement. Using a compass, I scribe and cut my adhesive-backed sanding discs as shown in the sketch. This gives me a number of coarse and fine rings and circles. I combine a coarse outer ring with a fine inner circle (or vice versa) to produce the dual-grit sanding capability. Depending on whether the fine grit is in the center or at the circumference, I find it necessary to change sander speeds to avoid burning the work, but this has not proved a drawback.

—*Gaylord R. Livingston, Chazy, N.Y.*

Belt sanding concave surfaces

Here's how to modify a belt sander to shape and smooth large-radius concave surfaces, such as a seat on a deacons' bench. Between the belt and the bottom of your sander, you will find a flexible, polished-steel plate that cuts down friction. Simply prepare a plywood shim with dimensions to fit your sander and one face curved to a slightly smaller radius than your work. Slip the shim under the flexible steel plate. Belt tension will pull the plate to the same curve as the shim, and will also hold the shim in place during operation. I easily sanded a curve with a 3-ft. radius this way, and don't see any reason why tighter curves couldn't be sanded as well. This method wouldn't work very well to dish out a bowl shape, however, as only the edges of the belt would cut.   —*Tom Hanson, Victor, Mont.*

Pattern sanding

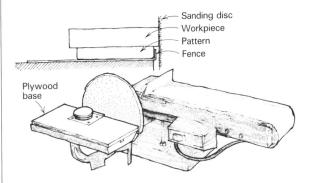

Sanding disc
Workpiece
Pattern
Fence

Plywood base

When I needed to reproduce 100 small ovals from ¼-in.-thick wood, I experimented first with several router methods, which proved either defect-prone or dangerous. I solved the problem by pattern sanding the ovals on a 12-in. disc sander. I suspect the same technique could be adapted to a belt sander with equally good results.

Only two parts are necessary, a pattern and a guide fence. I made a pattern from ¼-in.-thick Plexiglas, which I shaped and smoothed on the disc sander. Because the pattern rubs against the guide fence, it must be sized about ⅛ in. smaller than the dimensions of the finished work and should be chamfered on its bottom edge for dust clearance. The guide fence is simply a 4-in. square of sheet metal with one edge bent up to a 3/16-in. lip. Attach the guide fence to a plywood base with countersunk screws so that the lip overhangs one edge of the plywood slightly. Now clamp the plywood to the disc-sander table with the guide fence next to, but not touching the disc.

To use the device, fix the slightly oversize workpiece to the Plexiglas pattern. To keep the workpiece from slipping, use sandpaper, double-sided tape or protruding brad points. I've found that a sheet of sandpaper glued to the top of the pattern provides enough friction for most situations. Push the pattern against the sheet-metal fence and rotate it to grind the workpiece to shape. With 100-grit sandpaper, the whole operation takes about 30 seconds for a small, uncomplicated shape. The fixture shown may be used to form straight or convex shapes only. However, concave shapes could be easily cut using a similar device that incorporated a curved guide fence and small drum sander.   —*Don Herman, Brecksville, Ohio*

Production chamfering

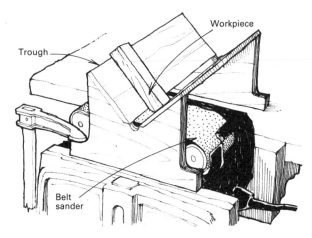

Trough
Workpiece
Belt sander

This setup helps you quickly sand an even chamfer on small parts. Build a trough with a slit in its bottom and position the trough straddling an inverted belt sander clamped in your workbench vise. The amount of chamfer is adjusted by raising or lowering the sander. Two words of caution: Don't obstruct the belt sander's ventilation opening when clamping the sander in the vise, and don't overtighten, lest you crack or distort the sander.   —*Fred Palmer, Pensacola, Fla.*

Stikit to the rescue

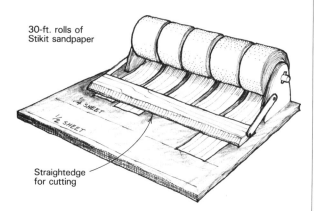

30-ft. rolls of
Stikit sandpaper

1/4 SHEET
1/2 SHEET

Straightedge
for cutting

I read Ben Erickson's letter expressing his displeasure with the Makita B04510 sander's thumb-torturing, paper-clamping arms (*FWW #62*). I also own a Makita sander, and agree with Erickson's low opinion of the clamping system. I just hope the garbage truck hasn't already hauled away his sander, because when the sturdy little machine is coupled with 3M's new Stikit paper system, it's a winner.

Stikit is a new adhesive-backed sandpaper that not only eliminates paper-clamping problems but also makes paper changing a snap. The product was developed expressly for orbital sanders, and is available in 30-ft. rolls in several grits, from 80 to 220. To use the paper, you first install a special pad on the bottom of your sander. Then, you simply press a sheet of the paper on the pad and you're ready to go. When the paper is worn, you just peel it off and put a fresh sheet on. The Stikit system not only makes changing paper easier—it also makes the sheets last longer since there are no bends to tear. Rolls of Stikit paper—along with special conversion pads that

can be applied to any orbital sander—are available from Trend-Lines, Inc. (375 Beacham St., Chelsea, Mass. 02150) or Woodworker's Supply of New Mexico (5604 Alameda N.E., Albuquerque, N.M. 87113).

To dispense the paper easily, I built a plywood and dowel rack like the one shown in the sketch. A fold-down straightedge of wood or metal holds the paper and provides an edge for cutting off the correct length with a utility knife.
—*Voicu Marian, Alliance, Ohio*

Quick edge sander

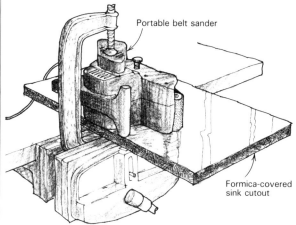

Portable belt sander

Formica-covered
sink cutout

Here's how to convert your belt sander to an edge sander quickly and easily. Clamp a piece of plywood (I use a Formica-covered sink cutout) to your belt sander with a large C-clamp, as shown in the sketch. Then tighten the clamp in your workbench vise.   —*Bob Elliott, Ankeny, Iowa*

Emery-cloth sanding spool

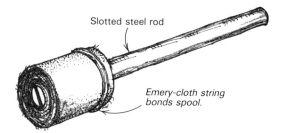

Slotted steel rod

Emery-cloth string
bonds spool.

I first discovered how to make these emery-cloth sanding spools 35 years ago, when I used them to finish steel forging dies. Start by hacksawing a slot in a 3⁄8-in. steel shaft. From a sheet of emery cloth, rip a ribbon slightly wider than the slot. Insert the ribbon in the slot and wrap by turning the shaft in the direction it will rotate. So far, what you've made is a flap sander, but there's an improvement.

From the sheet, rip a second ribbon, keeping this one stringlike, just about 6 or 8 threads wide. Wrap it around the lower part of the spool as shown in the sketch. Drop a little water on the string and, while holding a piece of scrap wood tight against the spool, turn on the drill to "burn in" the string. The glue in the threads of the string will bond tightly, banding the emery cloth into a firm spool that will last longer, sand smoother and won't flap all over your project.
—*Larry Stedman, Flushing, Mich.*

Sawing veneers on the bandsaw

To produce veneer slices on the bandsaw, we use a sharp blade, a tall rip fence aligned with the line of cut for each particular blade and an easily adjusted fingerboard rack. The fingerboard rack, which is the real secret to producing quality veneer, exerts even, soft pressure against the stock. This allows the sawyer to concentrate completely on pushing the stock smoothly and evenly through the blade. Because the cut is made in one continuous motion, you get a much cleaner piece of veneer. In fact, the veneer can usually be glued directly to the ground without having to thickness-sand errant sawmarks. Also, because the technique reduces the thick and thin spots in the slices, your veneer stock goes farther.

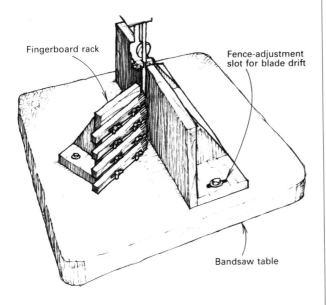

Fingerboard rack

Fence-adjustment slot for blade drift

Bandsaw table

Make the rack platform from ¾-in. plywood and the fingerboards from springy hardwood, such as ash or hickory. Cut sawkerfs in the end of the fingerboards both vertically and horizontally to soften the pressure, and fasten them to the platform with bolts through slotted holes. Leave the wingnuts on the back bolts loose so you can adjust and tighten the fingerboards easily with just the front wingnuts. With the fingerboards touching the veneer stock just in front of the blade, angle the rack about 60° from the line of cut and clamp the rack in place. If you intend to do much veneering, you may wish to drill and tap holes in your bandsaw table for mounting the fixture.

—*Jeff Simon and Mark Darlington, Steamboat Springs, Colo.*

Shop-built bandsaw fence

When I purchased my Delta 14-in. bandsaw, I decided not to buy the factory rip fence, because it clamps to the saw table only at one end. I do lots of hardwood resawing and believe that an accurate resaw requires a rigid fence clamped to the saw table at both the front and back.

Using standard hardware-store items, pieces from a tempered steel bed frame and lengths of 1½-in. cold-rolled square-steel tubing, I assembled an extremely sturdy, quickly adjustable fence for under $15. The fence's construction doesn't require

any welding, so it can be built and installed in an evening or two. The design allows the fence to be used on either side of the blade and to be angled slightly, as needed, to match the

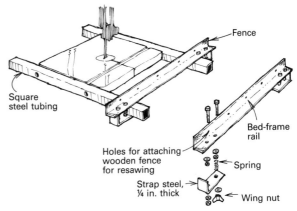

Fence

Square steel tubing

Holes for attaching wooden fence for resawing

Bed-frame rail

Spring

Strap steel, ¼ in. thick

Wing nut

blade's natural line of cut. I screw a taller wooden fence to the angle iron for resawing wide boards. Construction dimensions may be readily altered to suit the maker's saw.

—*Donald G. Sterchi, Bowling Green, Ky.*

Mitering trim on the bandsaw

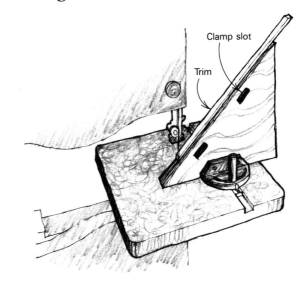

Clamp slot

Trim

When I needed to miter a few dozen ⅛-in.-thick cock beads, I first tried a chop box, but found it produced severe chipping behind the cut. So I built the fixture shown here to cut accurate miters on the bandsaw and avoid chipping. The fixture is simply a couple of 45° plywood triangles glued together and mounted upright to the saw's miter gauge. I routed two slots, as shown, to accommodate spring clamps that hold the trim to the fixture. Before use, push the fixture into the blade to leave a sawkerf halfway through. This sawkerf will serve as an index line for exact placement of the trim to be cut.

To use the fixture, first cut a trial piece to make sure the miter angle is an accurate 45°. If not, adjust the angle by tilting the saw table slightly. Next, cut each strip of trim about ½ in. longer than needed, then snip the mitered corners off in the fixture.

—*Dave Evenson, Cumberland, Wis.*

Sawing duplicate pieces on the bandsaw

I had experimented with various bandsaw techniques for sawing duplicate pieces, but I was dissatisfied with the results and the lack of flexibility. As a result, I developed this simple fixture. It lets you duplicate a pattern quickly and accurately, provided the curves aren't too abrupt for the blade to follow. Another advantage of the fixture is that you need only one pattern to produce various-sized duplicates.

The fixture evolved from a common circle-cutting jig that uses an auxiliary table clamped to a bandsaw table. The table is simply a piece of ¾-in. Formica-covered plywood with a ¾-in. groove routed in the top (the centerline of the groove is aimed at the cutting edge of the blade). The rest of the fixture consists of a guide pin, a sliding pivot pin and a spacer. The guide pin and the sliding pivot pin are both ¼-in. dowels pressed into small blocks sized to fit the groove.

To use the fixture, first cut a pattern an inch or two smaller than the finished size of your workpiece and drill a pivot-pin hole near its center. Next, place the guide pin in the groove at an appropriate distance from the blade. The distance from the guide pin to the blade will determine the enlargement of the finished duplicate over the pattern. Fix the guide pin at this distance by placing one or more spacer blocks between the guide and the end of the groove.

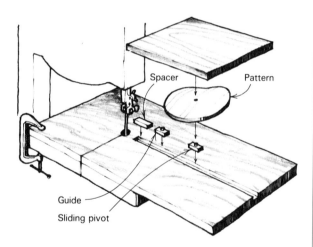

Now, attach a blank to the pattern with brads, two-sided tape or hot-melt glue. Fit the blank/pattern assembly over the pivot pin and cut the duplicate by rotating it while applying gentle pressure toward the guide pin. The pivot pin will slide back and forth in the groove as the pattern follows the guide pin. Generally, the stock is too large to allow the pattern to contact the guide pin at first. In this case, simply turn the pattern and feed the stock into the blade until the pattern touches the pin.

The fixture is useful for many applications. Curved pieces—such as the back of a crowned chair—can be ripped quickly and easily. If the fixture is adapted to a scroll saw, it can be used to cut elliptical picture frames. With minor modifications, the setup can also be used to cut circles.

—*Peter J. Cranford, Windsor Junction, Nova Scotia*

Clearing jigsaw sawdust with a flit gun

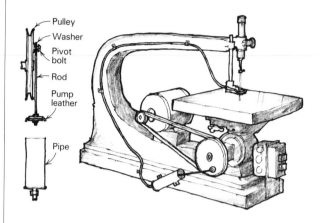

I've seen a couple of methods recently for clearing jigsaw sawdust with a hair dryer and with a vacuum. The "flit gun" method I have been using for 15 years does the job very effectively. The device doesn't consume much energy, is practically noiseless and produces short puffs of air to clear only the immediate area on the workpiece.

To build the flit gun, start with a length of 1½-in. brass sink drain pipe. I soldered a brass plate with a compression fitting to one end of the pipe, as shown in the sketch, but a wooden plug epoxied into the end of the pipe would work just as well. Now make up a plunger with a pump leather on one end and a washer for a crank pivot on the other. Locate a crank bolt about 2 in. from the center of the jigsaw's pulley. The gun's cylinder can be fastened at any convenient location on the jigsaw, either horizontally or vertically. Run plastic tubing from the end of the flit gun to a spot above the worktable. Turn on the jigsaw and let the flit gun fire away at the sawdust.

—*Edward J. Daly, Wyckoff, N.J.*

Cutting circles with a jigsaw

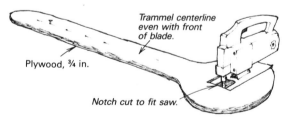

A jigsaw can quickly cut circles or large holes with this simple trammel made from scrap plywood. Cut the trammel leg to any convenient length, and make the head large enough so that you can saw a notch in it to seat the jigsaw. With the saw in place, draw a line perpendicular to the front edge of the blade and extend the line down the trammel's leg. Drill a pivot hole at the appropriate distance down the line, using a finish nail for a pivot pin. If you are going to save the circle you are cutting out, the kerf left by the saw must be outside the circle's circumference. The length from the center of the pivot pin to the side of the blade closest to the pin equals the circle's radius. If you are cutting a hole (the cutout will be scrap), the kerf must be inside the hole's circumference. In both cases, apply pressure downward and outward while cutting.

—*William S. White, Longwood, Fla.*

Jigsaw vacuum attachment

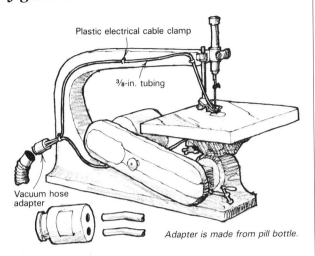

Adapter is made from pill bottle.

In past "Methods of Work" columns I have noticed a couple of ideas for blowing sawdust away from the cutting line on scroll saws or jigsaws. These methods seem to me to be contrary to the usual thinking about dust collection. On other machines dust is vacuumed, not blown around the shop to light in eyes, noses and motor bearings.

I've solved this problem with a few feet of ⅜-in. flexible plastic tubing, plastic electrical cable clamps, an empty plastic aspirin bottle and my shop vacuum. The aspirin bottle just happens to fit the adapter on the end of my vacuum hose. I attached the tubing to the aspirin bottle with clear silicone sealer. You can form the end of the plastic tubing to point it right where it is needed: Dip the tubing in boiling water for a few seconds and hold it in the shape you want while it cools. This simple arrangement works quite effectively to pull dust away from the work and out of my environment. I plug the saw and vacuum into a foot switch so they operate together.

—*R. J. West, Kansas City, Mo.*

Aquarium pump clears sawdust

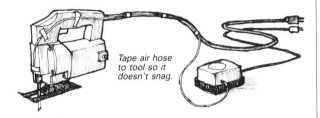

Tape air hose to tool so it doesn't snag.

Small diaphragm-type aquarium air pumps will supply a jet of air to keep sawdust away from pattern lines when scroll-sawing and the like. Fit the pump with a length of plastic tubing and tape the tube in place on the tool, aimed so it blows away the dust. The small pumps, which cost less than a good router bit, can be purchased at any aquarium supply store. Heavier pumps are also available and would serve with larger tools.

—*Michael H. Marcus, Portland, Ore.*

Jointing with a circular saw

Faced with the problem of jointing two 2x10 lengths of sugar pine for a carved sign, and not having a jointer, I used this method to produce an almost perfect joint.

First I bolted an 18-in. oak rail and a spacer of equal thickness to the base of my circular saw, as shown in the sketch. The rail acts as an extension to the base, ensuring the blade will run parallel to a straight fence. I replaced the combination blade with a hollow-ground planer blade.

Next I clamped the two 2x10s edge-to-edge on a level surface and clamped a hardwood fence on top of one of the planks. The fence was set so that the saw would pass right down the meeting line of the planks, removing some material from each plank.

After each pass of the blade I pushed the two planks together, readjusting the fence as necessary, and took another cut. After five or six passes, the joint was ready to be glued.

—*Robert P. Cromwell, Royalston, Mass.*

Lowering a radial-arm saw

Here is a dead-simple method for quickly and precisely lowering the blade of a radial-arm saw just enough for through-cuts. Lay a playing card (which you keep conveniently on top of the arm assembly) on the saw table over the saw's line of travel. Lower the blade onto the card until it buckles slightly but obviously under the pressure of the blade. The blade is then just at the correct level for through-cutting.

I invented this little trick after lowering the blade into the table with a thud several times, which I figured wasn't doing any good to the blade, the lowering mechanism or the saw's alignment.

—*Raymond Francis, Pelham, N.Y.*

Measuring stop block for power miter box

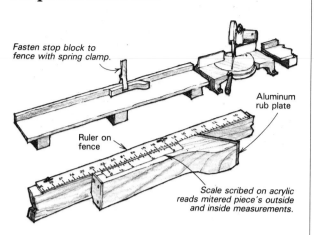

Fasten stop block to fence with spring clamp.

Aluminum rub plate

Ruler on fence

Scale scribed on acrylic reads mitered piece's outside and inside measurements.

If you cut lots of 45° miters on a power miter box, this device will save you hours of measuring. The scale on the stop block lets you set up quickly for either inside or outside measurements on a mitered frame.

To make the stop block, carefully miter a hardwood 1x2 and bandsaw to the shape shown. For a touch of class, install a 1/8-in.-thick aluminum rub plate to the face of the block. Screw a 6-in. length of acrylic to the top of the block and carefully scribe a measurement scale into the plastic. The measurement scale should be laid out in inches, but the numbering halved so that 1 in. is marked 1/2 and 2 in. is marked 1, etc. This scale will be used to set inside measurements as explained below.

If your saw doesn't have an inch scale along the fence, mount a metal yardstick to the top of the fence. Use slotted holes so you can fine-tune the position of the scale to reflect exact measurements.

To use the stop block for outside measurement miter cuts, simply align the zero mark on the stop block with the appropriate outside measurement on the fence scale. To use the block for inside measurements, first measure the width of your molding. Find the mark on the stop block that corresponds to the width and align that mark with the inside measure on the fence scale. For example, if your molding is 2-in. wide and the desired inside measure is 19 in., then find 2 on the stop block and align it with 19 on the fence side.

For added convenience, construct the stop block with a square end so it can be flipped and used for cutting pieces with square ends. Scribe a mark on another side of the stop block to align the square end with the fence scale. In use, fasten the stop block to the fence with a spring clamp or a small C-clamp.

—*Dean French, Kapaa, Kauai, Hawaii*

Odds & Ends

Chapter 14

Salvaging boards with loose knots

To salvage a board with an interesting but loose knot, fill the space around the knot with acrylic casting resin (available at hobby shops). First put tape on the back side of the knot and, with the taped side down, pour acrylic into the knot until it mounds up on top. After the resin has cured, you can scrape or sand it flush. Any scratches will disappear under a coat of lacquer or varnish.

—*David W. Worden, Pontiac, Mich.*

Clothes-iron shop applications

An ordinary clothes iron can simplify two furniture repairs: raising dents and reattaching loose veneer. To use the iron to remove dents and dings, set the heat to "cotton" or "wool," wet a cotton cloth pad and place it over the dent. Press the iron to the pad for two or three seconds and check your progress. Repeat the procedure until the dent is flush.

To repair loose veneer, place a damp cloth between the iron and the work. Apply the iron to the spot, taking care to move it about so as not build up the heat too fast—the veneer will scorch if you're not careful. This method doesn't work with some adhesives, but most old furniture was veneered with hide glue, which will reactivate and hold the loose veneer down tight again.

—*Rollie Johnson, Sauk Rapids, Minn.*

Homemade lock screws

Nylon-insert lock screws are quite effective when you need a bolt that won't loosen in vibrating machinery. But the commercial versions are not available in every size and are expensive. To make your own, simply drill a hole through the bolt near the end and insert a short length of heavy-gauge nylon cord of the type used in grass trimmers. Trim the cord flush with the threads and screw the bolt into the hole. The nylon will crush into a form fit of the threads and will hold beautifully.

—*Gordie Mulholland, Streator, Ill.*

Lubricating sealed bearings

If you ever have a power tool with a noisy sealed bearing, here's an unorthodox but effective way to lubricate it. Place the dry bearing in a can of oil, and place the can inside a bell jar on a vacuum pump. Switch the pump on and watch the air bubbles come to the surface of the oil. When the bubbles stop rising, turn the pump off. Normal air pressure will force the oil into the sealed bearing.

Some of you are probably asking where you can get a vacuum pump and bell jar. Check with the head of your local high school's science department. If you're a tax-paying resident of that town, you'll most likely be allowed to use the school's equipment. —*William Warner, York, Pa.*

Wired tambours

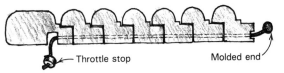

Throttle stop — Molded end

Last year one of the students in my high-school woodworking class made a roll-top desk based on Dale Tucker's wired tambours article *(FWW #48)*. As a substitute for the vinyl-coated stainless-steel cable Tucker recommended for stringing together the slats, we used ordinary bicycle brake cable, which is easy to find, strong, flexible and comes with a ready-made stop molded onto one end.

To make the tamboured desk top, we first shaped and cut slats with the profile shown. To align the wire holes in each slat we drilled holes in the first slat, clamped it to the benchtop and used it to guide the drill for the other slats. After all the holes were drilled, we threaded the bicycle cable through the slats, pulled it tight and locked it with a throttle stop from the local auto-parts store.

—*Sam Gardner, Duncan, Ariz.*

Box-lid trick

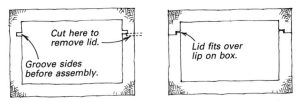

Cut here to remove lid. — Groove sides before assembly.

Lid fits over lip on box.

When I make small boxes, I assemble them in one piece, including the lid. Then when I cut the lid off, it will match the box exactly. When I want an undercut on the lid to fit over a lip on the box, I use a variation of the one-piece technique. First, before assembly, I cut a 1/8-in.-wide groove in the inside face of each of the box's sides where the top of the lip is to be. After the box has been assembled, I use a narrower blade to cut off the lid. I offset the second lid-removal cut from the inside groove, as shown, to produce a lip that nests into the lid.

—*F. B. Woestemeyer, West Chester, Pa.*

Two hidden shelf hangers

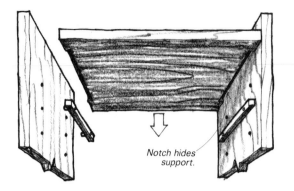

Notch hides support.

Here's how I provide adjustable shelves in bookcases and hide the unattractive support hardware. First I cut a ⅜-in. thick, ½-in. wide support slightly shorter than the width of the shelf. I peg the support with two ¼-in. or ⅜-in. dowels and drill a series of holes on each side of the carcase to match the pegs. Finally I rout a recess into each end of the shelf to accommodate the support so that it is visible only from below.   —*Alan Platt, LaGrangeville, N.Y.*

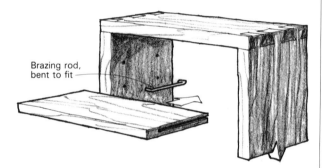

Brazing rod, bent to fit

I use these hidden shelf-supports in small wall cabinets and the like. I suppose they could be scaled up for larger carcases, but they seem better suited for smaller work with, say, ½-in. thick, 18-in. long shelves.

First bend brass wire into wide U-shapes, as shown in the sketch. I like to use ³⁄₁₆-in. diameter brazing rod. The base of the U must be the same length in all the pieces. I use the jaw width of a small machinists' vise as a handy length gauge. Heat the rod with a propane torch if you have trouble with the rod breaking at the corners.

Next drill a series of paired holes along both sides of the cabinet to fit your brass-rod pins. Cut a ³⁄₁₆-in. stopped groove down the center of each end of your shelf. To install the shelf simply pop two brass supports into matching holes and slide the shelf on from the front. The support is locked in place and is perfectly invisible.

—*Fred Gati, Providence, R.I.*

All-wood adjustable shelf bracket

This easy-to-make shelf bracket ensures accuracy because both pairs of shelf-height notches are established with one hole. To make the bracket, clamp two 1x2 strips together and drill a

series of holes down the centerline through both strips. The holes will set the spacing between shelf locations. Now rip each 1x2 on its centerline to produce two matching brackets for each

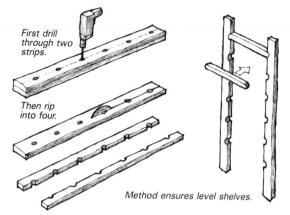

First drill through two strips.

Then rip into four.

Method ensures level shelves.

end of the shelf unit. Install the brackets in the carcase and cut several ¾-in. square shelf supports to fit in the notches. Round the ends of the shelf supports to match the half-round notches in the brackets.   —*Rollie Johnson, Sauk Rapids, Minn.*

Assembly squares

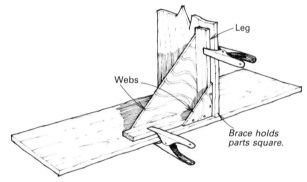

Leg

Webs

Brace holds parts square.

For assembling cabinets by myself, I have a set of assembly squares that I spring-clamp into corners to hold the parts perpendicular to each other. The squares act as a second pair of hands, holding the workpieces square and in alignment until I can spread the glue or drive home the screws. The webs are ¼-in. plywood, and the legs are 1x2s.

—*T.D. Culver, Cleveland Heights, Ohio*

Velvet drawer bottoms

The standard approach to lining the bottom of jewelry-box drawers is to cut a piece of felt to size and glue it in place. I prefer velvet's rich feel over felt, but the cut-and-paste approach doesn't work as well with this material.

To solve the problem, I install the velvet bottom before assembling the drawer. I apply a light, even film of Titebond or Elmer's glue on an oversized plywood drawer bottom. I lay the velvet on the glue-wetted surface, smooth out the wrinkles and—when the glue has dried—cut the bottom to size. If you cut the bottom to size before the glue sets, the wet threads in the velvet tangle in the saw and create a mess.

The rest of the assembly is as usual, except you need to cut the bottom grooves in the sides of the drawer wider by ¹⁄₁₆ in. or so to accommodate the extra material.

—*David Miller, Annville, Pa.*

Velvet drawer bottoms revisited

Frankly, I think David Miller is working too hard. There's another method for installing velvet drawer bottoms and jewelry box linings, using upholstery techniques. The result looks better, allows replacement and can be adapted to the sides and top of the box as well. First cut a piece of thin cardboard slightly smaller than the bottom of the drawer, then cut a piece of velvet a little longer and wider than the cardboard. With the velvet face-down on a table, center the cardboard on the velvet and trim each corner of the velvet at 45°. Apply a bead of quick-drying glue to one edge of the cardboard and fold the velvet's seam into the glue. After the first edge has set for a few minutes, glue the other edges to the cardboard, stretching the velvet as you go so there are no wrinkles on the face side. Finally, apply a dab of glue to the center of the cardboard back and press the bottom into place in the box.

This technique works well with velvet, felt and leather. It even works with silks and sateens, which telegraph glue spots badly and thus can't be glued directly to wood. This approach also works well with shadowboxes and collection displays, because you can mount the display items to the velvet insert with fine wire before placing the insert in the box.

—*Ernest B. Shipley, Oakland, Calif.*

Blocks for squaring a carcase

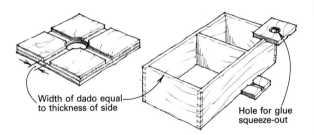

Width of dado equal to thickness of side

Hole for glue squeeze-out

While gluing up and clamping a cabinet case without a back, it was difficult to keep the case square. So, I made the alignment blocks shown above by cross-dadoing an 8-in. square of ¾-in. plywood. The width of the dado is equal to the thickness of the case material. I drilled the center of the block to prevent glue squeeze-out from permanently attaching the blocks to the case.

—*Gaylord R. Livingston, Chazy, N.Y.*

Making a shop moisture gauge

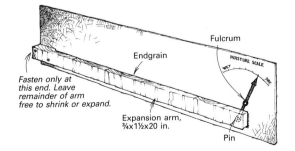

Fulcrum

Endgrain

MOISTURE SCALE
WET   DRY

Fasten only at this end. Leave remainder of arm free to shrink or expand.

Expansion arm, ¾x1½x20 in.

Pin

The sketch above shows a simple gauge you can make that will give you a general idea of the relative humidity in your shop. The gauge will graphically show your customers how wood moves and why you build the way you do. For the gauge to work properly, the 20-in. wooden expansion arm must be

sliced off the end of a wide panel or glued up from flatsawn segments, as shown. Movement of the gauge will be more dramatic if you pick a wood species that has a large tangential shrinkage percentage, such as beech, sugar maple or white oak. If you have access to a moisture meter, you can scale the gauge numerically.

—*John Sillick, Gasport, N.Y.*

Clock-cavity routing jig

Router template hole

Hinged template arm

Nuts for height adjustment

Eyebolt

Foam pad

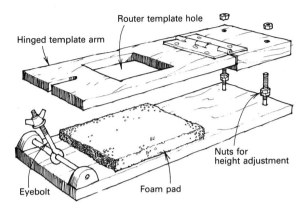

Although I devised this jig for routing a cavity for a quartz clock works, the idea could be adapted to many routing operations. The beauty of the jig is that it combines a workpiece hold-down and a routing template into one device. It's well suited for small production runs because you can quickly pop workpieces in and out of the jig.

The jig consists of a base, a foam pad, the hinged template arm and an eye-bolt. The eye-bolt pivots on a bolt axle to fasten the template arm down over the workpiece. I attached the template arm to the base with bolts as shown in the sketch. The template arm can be set for the thickness of the workpiece by adjusting the nuts below the arm.

To use the jig, simply lay the workpiece on the foam pad, lower the template arm, lock it down with the wing nut and rout away.

—*Les Stern, Denver, Colo.*

Screw plugs on a strip

Waste

Stock

Tape holds plugs.

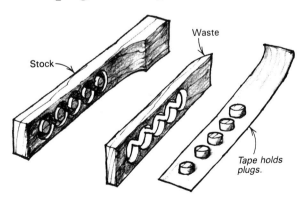

After using the plug cutter to cut a row of shallow plugs for screw hole covers, tape the row with masking tape and rip as required with the bandsaw. The result, as shown in the sketch above, is a neat strip of shallow plugs, all with the same grain direction, ready for installation right off the tape.

—*Robert M. Vaughan, Roanoke, Va.*

Fancy eyes for pull toys

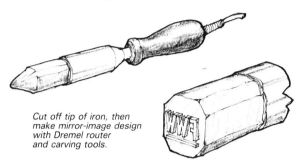

Cut plug from tube,
install in eye hole.

Drill pupil hole,
install brass rod.

Here's how to make eyes for wooden-animal pull toys that give them an animated and professional touch. First, drive a 3-in. dowel into a 3-in. length of brass tubing with the same outside diameter as you've chosen for the eye. If you round the nose of the dowel and proceed carefully, you should be able to drive the full length of the dowel through.

Next, cut a short plug off the brass-encased dowel and epoxy it into an eye hole you've drilled in the toy's head. When the glue has set, drill an off-center pupil hole near the bottom edge, and tap in a piece of brass rod. Use a file and sandpaper to bring the eye flush with the toy's head. A drop of lacquer over the entire eye will prevent the brass from tarnishing and preserve the gleam. Brass tubing and rods are available in various sizes at most hobby stores.

—*Mark DiBona, Kensington, N.H.*

Double bifocal shop glasses

Last year, it finally became necessary for me to begin wearing bifocals. The close-up area of the lens worked fine—as long as the work was below me. But the glasses didn't help at all if I was doing close or detailed work overhead. So, I had my optician make me a pair of shop glasses with bifocal areas on both the top and the bottom, as shown above. The particular design is known as Double D28, Occupational Segments. The glasses have greatly improved my enjoyment of woodworking.

—*Rod Goettelmann, Vincentown, N.J.*

Homemade wood-branding iron

If you have always wanted one of those fancy branding irons to mark your projects but felt they were too expensive, here's how to make one in your shop for next to nothing. First, scour the local flea markets to obtain a large electric soldering iron. Some of the older ones have copper tips a full inch across. The iron must work, but the condition of the tip is not important.

Cut about half the tip off to leave a large flat across the end of the iron. File this smooth. Trace your name or logo on the copper face, remembering that the design must be the mirror image of what you want to stamp on your projects. Rout around the letters to a depth of $\frac{1}{16}$ in. with a Dremel tool, then use a small chisel or your woodcarving tools to finish up the design and add crisp edges. Copper is soft and will pare away easily—like carving lilac end grain.

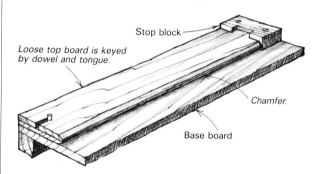

Cut off tip of iron, then
make mirror-image design
with Dremel router
and carving tools.

If you can't locate an old soldering iron of sufficient size to handle your design, an alternative approach is to flatten the tip as above, but carve the design on a separate chunk of solid copper. Then use a torch and high-temperature silver solder to attach the plate to the iron.

To use the branding iron just let it heat up and press the copper against the wood. Presto! Your name is permanently charred in wood. —*Wayne Spicer, Memramcook, N.B.*

Improved veneer-shooting board

Stop block

Loose top board is keyed
by dowel and tongue.

Chamfer.

Base board

Ian Kirby's design for a veneer-shooting board *(FWW #47)* reminded me of the modifications I have made to mine. Initially I wanted to make the device easier for school children to use, but I soon preferred the new model, too.

One of the difficulties with the original was lining up the loose top board with the baseboard without moving the veneer. My modifications hold the ends of the top board so that it remains aligned with the base. This makes it easy to slip veneers between the two boards into the correct position for planing. You could construct the shooting board with dowels at both ends instead of the keyed stop, but I use mine for shooting small panels, too—I simply remove the top board and square the panel against the stop for planing the edges.

—*Ernie Ives, Ipswich, England*

Patching veneer

This veneer-patching technique is not only easier than the "cut and fit" approach, it also results in a virtually invisible repair if the color and grain of the patch is matched carefully to the original veneer. First feather the edges of the missing veneer defect so that they taper, as shown at right. Select a piece of veneer for the patch slightly larger than the defect area and glue it into the recess. Then just scrape and sand the raised edges of the patch flush with the surrounding surface.

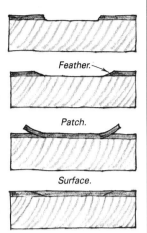

Feather.

Patch.

Surface.

—*Rollie Johnson,*
Sauk Rapids, Minn.

Heat-bending veneer strips

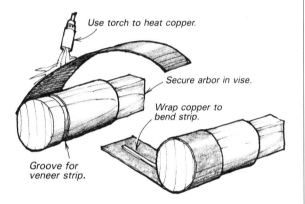

Use torch to heat copper.

Secure arbor in vise.

Wrap copper to bend strip.

Groove for veneer strip.

With this simple-to-build device you can bend strips of veneer for inlaying or edging. The chances of cracking or breaking the veneer are greatly reduced.

First turn or saw a wood arbor to a slightly smaller diameter than the bend needed. This tighter radius allows for a little springback after the veneer has been bent. Cut a square end on the arbor so it can be clamped in a vise. Fit the arbor with a thick (0.025-in. or thicker), wide copper strap that will retain plenty of heat. You can anchor the strap with tacks or by simply forcing it into a groove. Extend the strap a foot or more so the end will stay cool enough to handle without gloves.

Hold the copper strip away from the arbor and apply the torch, heating the strip well beyond the part that will touch the veneer to reduce the rate at which heat dissipates from the working area. When the strap is as hot as it can be without scorching the veneer, push one end of the veneer under the strap, pull the strap tight and wrap it around the arbor. Hold the strap in place for a minute while it cools.

—*Howard C. Lawrence, Cherry Hill, N.J.*

Wooden box hinge

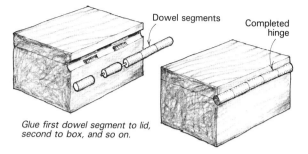

Dowel segments

Completed hinge

Glue first dowel segment to lid, second to box, and so on.

This wooden hinge is tricky to make. But because it's distinctive, attractive and functional as well, perhaps the extra trouble is justified.

To make the hinge, you'll need a ½-in. or ⅜-in. dowel, preferably from the same wood as your box, a brazing rod pin as long as the back of the box and a core-box router bit to rout a round-bottomed slot the same diameter as your dowel.

Start by determining the number of hinge segments you wish to have. There should be an odd number, and each segment should be no longer than 1½ in. or so. Divide the length of the box back by the number you have chosen to get the length of each dowel segment. Now carefully slice up the dowel, taking care that each segment's end is a perfect 90°. The next step is to drill a hinge-in hole through each of the interior segments and halfway through both of the end segments. I've seen several ideas in the "Methods of Work" column to accomplish this operation. The easiest, I think, is to clamp a 2x4 to the drill-press table and drill a registration hole the same diameter as the dowel about ¾ in. deep into the 2x4. Without moving the drill-press table, chuck the bit you intend to use for the pin holes into the drill press. When you place a dowel segment in the registration hole, the pin bit will be centered right over it.

Now you're ready to rout a round-bottomed channel centered over the seam where the top of the box meets the back. You can rout this channel by clamping the top to the box and using a core-box bit or, alternatively, you can rout the top and back separately with a piloted cove bit. Either technique will work.

When you have completed the hinge channel, you're ready for the tricky part—gluing the hinge in place. First string the dowel segments on the pin wire like beads, with an end segment on each end. Lay the hinge in the channel between back and top and mark the location of each segment. Carefully smear dabs of glue in the channel, alternating between top and back so that half of the segments will be glued to the top and half to the back. Don't use too much glue, because any squeeze-out will lock up the hinge. I've experimented with several types of glue and have had the best luck with epoxy, even though I normally avoid it. Place the hinge into the channel and clamp lightly to minimize squeeze-out. When the glue sets, remove the clamp, cross your fingers and try the hinge action. If all has gone well, you will have a smooth-working, good-looking hinge.

—*Jeris Chamey, Ponca City, Okla.*

Built-in table extension

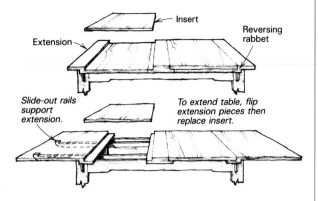

Insert

Extension

Reversing rabbet

Slide-out rails support extension.

To extend table, flip extension pieces then replace insert.

Here's a table that can be extended without the nuisance of separately stored leaves. The table's top consists of a fastened central piece, two inserts and two extensions. To extend the table, lift up the inserts, then flip the extensions so the rabbets under their ends are facing up. The two inserts fit into the rabbets, and two rails slide out from the table's ends to support the extensions.

—*Brian Tinius, North Hollywood, Calif.*

Wall hanger hardware

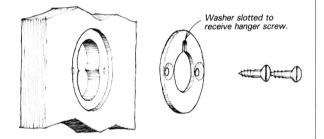

Washer slotted to receive hanger screw.

This modified 7/8-in.-OD steel washer lets you hang shelf brackets or wall cabinets flush against the wall. Drill a 9/64-in. hole near the inside edge of the washer and file the space between the hole and the washer opening to produce a slot. Now drill two countersunk holes in the washer for mounting screws. For lighter applications, you can skip the mounting screws and epoxy the washer in the recess. To install the hanger, drill a shallow 7/8-in.-dia. recess in the workpiece so the washer can be screwed flush to the surface. Drill two stopped, overlapping 3/8-in.-dia. holes in the workpiece to make an oval-like cavity under the washer. The cavity allows for the downward movement of the cabinet or bracket over the head of the hanger screw, which is driven into a wall stud.

—*Robert W. Terry, Palm Beach, Fla.*

Wooden drawer-pulls

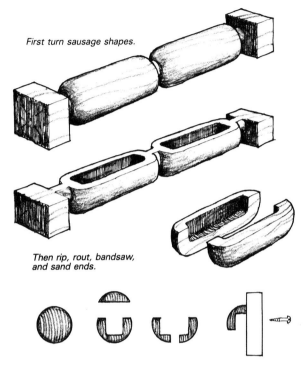

First turn sausage shapes.

Then rip, rout, bandsaw, and sand ends.

When I couldn't find any solid-oak drawer-pulls that I liked, I devised this method to make my own. To make four pulls, mount a 12-in.-long, 2-in.-square blank on your lathe. Turn two 4¼-in.-long, 1⅞-in.-dia. sausages. Then rip or plane ½ in. off one side of the sausages to produce a flat face. Rout hollows in the flat with a core-box bit as shown in the sketch. To complete, split the blank lengthwise with a bandsaw, cut the rough pulls apart and finish the rounded ends with a disc sander. Install the pulls with two screws—one each into the solid wood on both ends.

—*Gary P. Korneman, St. Joseph, Mo.*

INDEX

Index

Editor: Andrew Schultz
Copy/production editor: Victoria W. Monks
Computer applications specialist: Dinah A. George
Print production manager: Peggy Dutton

Typeface: ITC Garamond
Paper: Marcy Matte, 70 lb., neutral pH
Printer and binder: Arcata Graphics, Kingsport, Tennessee